Praise for *When a Parent Has Cancer*

"For any parent with cancer, our children become the focal point of our greatest fear—dying before we raise them—and our greatest emotional turmoil as they try to grow up amid the exhaustion, pain, and joy of living with the 'elephant in the living room'—cancer. Wendy Harpham has written the guidebook that will become every parent's answered prayer."

—KATHY LaTOUR, author of *The Breast Cancer Companion*

"With the glut of books on the market to advise and counsel people living with a diagnosis of cancer, few have come close to capturing the realistic experience of being a parent with cancer and the extraordinary implications that this diagnosis has for children they love. With this book and its companion guide for children, Wendy Harpham has given all parents diagnosed with cancer bountiful, practical advice on how to cope and even thrive in their relationship with their children. As she did with her other wonderful books on living with cancer, Wendy has bestowed a gift to all of us with this exceptional publication."

—ELLEN STOVALL, Executive Director,
National Coalition for Cancer Survivorship

"I cannot think of a person better qualified to write a book advising parents who have cancer on how to deal with their children than Wendy Harpham. The language of the book is so easy to understand that it is likely that the reader will be unaware of just how much instruction he or she is getting. I heartily recommend this book for those cancer patients who seek to maintain the best possible relationship with their children. The novel idea of including a young girl's perspective on the effect of her mother's cancer on the little girl's life is a wonderful concept. Parent and child learn together. An unbeatable combination."

—HAROLD H. BENJAMIN, PH.D., Founder and President,
the Wellness Community

"I was captivated by this survivorship story, by Dr. Harpham's commitment to this labor of love, and the incredible gift she has created for other survivors and families out of her own complex cancer experience. The message that cancer need not rob kids of their childhood but can actually contribute in very positive ways to coping skills that will serve them well through a lifetime is especially meaningful. I found the text to be realistic *and* hopeful, encouraging, and upbeat. A valuable resource, especially for people who lack access to the specialized doctors, nurses, social workers, and counselors who might help lessen the burdens associated with the cancer experience."

—P. J. HAYLOCK, Cancer Care Consultant and President, Oncology Nursing Society

"*When a Parent Has Cancer* provides comfort, reassurance, and practical advice to families facing cancer. The combination of the parent guide and the children's story, *Becky and the Worry Cup*, is extraordinarily effective."

—DIANE BLUM, M.S.W., Executive Director, Cancer Care, Inc.

"As a pediatrician, I say 'Thank you, Dr. Harpham for your lovely prose, urgent message, and practical advice to parents with cancer.' While the physician needs to discuss the problems generated by this medical nightmare, *When a Parent Has Cancer* will make a very difficult task much more humane and understandable for parents and their medical team. The companion children's book, *Becky and the Worry Cup*, is a stroke of genius, showing families how to deal with the reality of a parent's cancer in an honest, helpful, and life-enhancing way."

—RUTH LAWRENCE, M.D., Professor of Pediatrics, Obstetrics, and Gynecology, University of Rochester Medical Center

"Among the hardest questions posed to me in helping patients and their families face cancer is 'What do I tell my child?' Wendy Harpham has answered those questions from her own experience. Uniquely, however, she has taken the next step—to provide other parents with a guide that should help them to talk more comfortably with their children, and to find the words with which to discuss this subject that frightens just by its name alone. Wendy has steadfastly examined her emotions and reactions with an honesty and sincerity that can only bring reassurance and hope to parents who are facing this crisis for the first time."

—JIMMIE HOLLAND, M.D., Chief, Psychiatry Services, Memorial Sloan-Kettering Cancer Center

"For all of us struggling to find the way through the maze and tailspins of cancer, there is hope and very clear direction in the experiences and words of this book. Through honest, gutsy, yet innocent windows, Dr. Harpham lets us see that 'happily ever after' can become 'happier even after.' She has taken the chaotic cancer challenge and translated it into words of calm reassurance and stable guidance so others' challenges will be lessened. What a gift!"

—JUDY GERNER, Director, Anderson Network, U.T.M.D. Anderson Cancer Center

"A sensible and sensitive guide to helping children through cancer that threatens a parent's life. Dr. Harpham is knowledgeable, clear, and kind. She gives advice that is useful and easy to understand. The story written for children is a real bonus."

—DAVID SPIEGEL, M.D., Professor of Psychiatry and Behavioral Sciences, Stanford University School of Medicine

Gail Nogle

About the Author

WENDY SCHLESSEL HARPHAM, M.D., is an internist in Dallas, Texas, where she lives with her husband and three children. She is the author of *After Cancer: A Guide to Your New Life, Diagnosis Cancer: Your Guide Through the First Few Months,* and *The Hope Tree* (coauthored with Laura Numeroff).

The Harpham Family: (left to right) Rebecca, Jessica, Ted, Wendy, William

WHEN A PARENT HAS CANCER

A Guide to Caring for Your Children

Wendy Schlessel Harpham, M.D.

Perennial Currents
An Imprint of HarperCollins*Publishers*

AUTHOR'S NOTE

Becky and the Worry Cup and *When a Parent Has Cancer* are based on my personal experiences as a physician and cancer survivor. To protect the privacy of others, I have changed names and altered identifying characteristics. In some cases, events were modified and stories merged for the purpose of illustrating a point. This book is not intended to substitute for professional care. It serves to supplement the information provided by your doctors, nurses, and counselors.

A hardcover edition of this book was originally published in 1997 by HarperCollins Publishers.

First Perennial Currents edition published 2004.

Library of Congress Cataloging-in-Publication Data
Harpham, Wendy Schlessel.
When a parent has cancer : a guide to caring for your children / Wendy Schlessel Harpham.— 1st Perennial Currents ed.
p. cm.
Previously published: New York, NY : HarperCollins, c1997.
Accompanied by: Becky and the worry cup / story by Wendy S. Harpham ; illustration by Jonas Kulikauskas.
ISBN 0-06-074081-7 (pbk. :alk. paper)
1. Children of cancer patients. 2. Cancer—Patients—Family relationships. 3. Cancer—Psychological aspects. I. Harpham, Wendy Schlessel. Becky and the worry cup. II. Kulikauskas, Jonas. III. Title.

RC262.H295 2004
362.196'994—dc22

2004040058

05 06 07 08 ◆ /RRD 10 9 8 7 6 5 4 3 2

For My Children

REBECCA ANNE

JESSICA MARTHA

WILLIAM SAMUEL

Contents

Acknowledgments

Innumerable people have left their mark on these books. My older sister, Debra Sue Bruck, a graphic artist, drew illustrations for the earliest version of the children's book. Her devotion inspired my belief in the concept of an illustrated children's story, and provided much-needed distraction from my illness.

A few people deserve special thanks for offering detailed comments on many versions of both books, and acting as sounding boards when major changes in form or content were being considered: Clare Buie Chaney, Ph.D., Shauna Erickson, Jane Ramberg, Sharon Witherspoon, Jeff Gillman, Ph.D., Andrea Bedway, Ph.D., O.C.N., Mary Jacobs, Jean M. Harpham, and Margaret C. Sheperd. And, Kathleen McCue on Chapter 8.

Since the first draft in 1992, many friends and colleagues read one or more versions, or even just a section or two, and gave me feedback that helped shape a phrase or philosophy, or the tone of the work. I apologize for any unintended omissions: Adele M. O'Reilly, Anne Heideck, L.C.S.W., Lawrence Frenkel, M.D., Julia Rowland, Ph.D., Billie McAnally, M.D., Erin Tierney Kramp, Claude Prestidge, M.D., Jessica Miller, Barbra Silverman, M.S.W., Brenda Casey, R.N., O.C.N., Harold Y. Vanderpool, Ph.D., Mary Cunnane, Judy Levin, Wendy Kahn, Susan O'Steen Allen, Kathleen Keys, Jan O'Reilly, Elizabeth Henderson, L.C.S.W., Ray J. Stankey, Kimberly Colen, Debbie Wills, Roger R. Bick, M.D., Gloria Weiner, Paula Chabanais, Carol Bratcher, Karen Roos, Diane and Tami Gorsline, Judith L. Szatmary, Susan Creagan, M.D., Mike, Eleanor, Zack, and Ava Bastian, and my close friend whose untimely death intensified my drive to write and whose voice I still hear in my head, Ellen Hermanson, the loving mother of Leora.

Special thanks go to my editor, Roger Jones, who believed in this book set and made it a reality despite the unique challenges, and to Fiona Hallowell, editor, for her candid input on the text and illustrations. She deserves a gold medal for her patience with the jillion faxes and phone calls of my electronic hovering over the manuscript. And, sincere thanks to my new editor, Alison Callahan, and Cesar Garza for making possible the updated paperback edition.

Heartfelt thanks to Jonas Kulikauskas. His creative and whimsical illustrations capture the balance of seriousness and lightheartedness needed to make the story work well.

Others have left their imprint on these books through their influence on my parenting. I am indebted to my internist, Susan Creagan, M.D., my primary oncologist, James Strauss, M.D., and the oncologist who treated me in clinical trials, David G. Maloney, M.D., Ph.D. All three have shown ongoing concern for my family's well-being, and have maximized my ability to care for my children. Dr. Creagan has seen me through the worst days and nights, sometimes sacrificing her own family time in order to do so. I am grateful not only for her invaluable insights about being an ill parent and for her wise advice but for knowing best when I needed her words as my physician, colleague, or friend. I owe thanks to the many other physicians and nurses who have been involved in my care over the past six years, all of whom have worked hard to speed my recovery.

Claude Prestidge, M.D., my children's pediatrician, has helped me enormously. In addition to caring for my children, he has responded to all of my concerns, both rational and silly, with expertise and sensitivity.

Rabbi Jeffrey Leynor has continued to be a source of comfort, strength, and guidance as I have tackled existential questions and have tried to nourish my children's spirituality.

Pam Jenkins, M.S.W., a truly gifted oncology social worker, has also continued to be a vital source of comfort, strength, and guidance but in a very different way. Counseling with Pam has been the hardest work I've ever done. In large part because of her, my "new normal" is a better one.

I can never fully thank all the friends, family, colleagues, acquaintances, and strangers who have pitched in to help my family. For the first half of 1991, meals, babysitting, and carpooling were generously provided by families of Congregation Beth Torah, and a group of doctors and nurses at Presbyterian Hospital of Dallas. Family and longtime friends visited from out of town to help care for the children when I couldn't. Ted's parents, Jean and John, moved into our home to care for our children when I was first sick, and when Ted and I went to California in 1993. Friends and family helped Ted with the children whenever I traveled by myself for treatment. My parents, Joseph and Naomi, and my siblings, Richard, Debra, and Donna, provided love and support.

It is hard on families when a parent has to be treated far from home. An anonymous group of my colleagues gave my family an invaluable gift by making it possible for me to travel back and forth each week between treatments instead of staying in California for the entire treatment period. Roberta and Harry Forbes drove me to and from the airport, thus making goodbyes with my children and Ted less painful. Tina Marie Liles, R.N., not only ensured that my medical care went smoothly at Stanford University Medical Center, she helped me through the separations from my family and the stresses inherent in clinical trials. Each time I was admitted to Stanford, the nurses and social workers extended themselves to me. I met generous people like Ezra Soloman and Bea Fioretti, who took me under their wings, providing transportation and companionship whenever I was away from home. The parents of my children's teammates from the various sports cheered my children in my absence. When I was able to attend the games, between bleacher discussions of headers and passes, screens and rebounds, set-ups and serves, and tags and bunts, these parents listened to me talk about everything from monoclonal antibodies and chemotherapy to the twenty-seventh possible title for this book. And, there are many, many more people who have offered themselves in whatever way might be helpful. To all of you, thank you.

The most important person in my life has made this book possible—my husband, Ted. Everything I've survived, we've survived together: cancer, six courses of cancer treatment, closing my practice, and soccer season. He has read and commented on countless drafts of both books, with his genius for organization and tone, and the insights that only he could provide. Ted is a private person; his offer to share our story reflects his selfless dedication to this project.

My three children, Rebecca Anne, Jessica Martha, and William Samuel, have been my inspiration. Their candid curiosity has driven me to find honest answers; their youthful dreams have pushed me to find and nourish hope; and their trust has forced me to find courage. They've offered comments on dozens of versions of the children's book and their feedback has been an essential factor in determining the direction and details of the illustrations. Their willingness to share some of our private moments in order to help other families makes me proud and grateful. As we all saw on their baseball fields this summer, you can be down in the bottom of the ninth, with two outs and two strikes, and still hit a home run to win the game.

Important Message to Parents, Friends, and Extended Family

This book discusses the difficulties of raising children when a parent has cancer and proposes approaches for preventing and responding to common problems in a healthy way. Addressed to the parent with cancer, *When a Parent Has Cancer* also can help healthy parents whose child's grandparent or teacher has been diagnosed. In addition, the information is intended for the other significant adults (grandparents, aunts, uncles, godparents, teachers, coaches, and family friends) in the lives of children. There is one caveat: Only if readers understand the distinct place of each person in the children's lives can this book help to heal the children.

For Parents: My Family—Your Family

Someone important to your children has cancer. The patient may be you, your spouse, or your children's grandparent or teacher. You are trying to figure out how to integrate this new experience into your family's life. The text is addressed to you because, in most cases, you are the legal guardians and primary caregivers; raising the children is of greatest immediacy to you. And, we are talking about your children. It is your family's life.

Raising children is really hard. There is no one "right" way to do it. All you can do is your best, as you see it at the time. Friends, extended family, and even professionals may think they know what's better for your children. But they are all looking in from the outside. And even though I am a parent surviving cancer, I've only been on the inside of my family's experience, not yours.

Your family has its own style and traditions, emotional makeup, history, and strengths and weaknesses. In the prologue that follows, I will tell you how I depended on friends and family. This was included not only to confirm the value to me of others' support but to encourage you to ask yourself questions like, "Do *I* need more help at home?" and "To whom do *I* turn for help?" I will share with you a few deeply personal conversations I had with my oldest child, Rebecca, not so that you could learn about *her* fears and feelings but to stir questions in your mind about *your* children: "What are *my* children's fears and feelings?" and "How do I want to respond to them?" Discoveries such as the worry cup and Mad-Busters in *Becky and the Worry Cup* are not described as the only ways to resolve your children's worries and anger, but as the cue for you to ask, "What do *we* do with *our* worries and anger?" Your answers will guide you in caring for your children.

The family-centered approach I take here is intended as encouragement to be true to yourself while you raise your children. Do what is appropriate for you and your family. Like all parents, you'll make mistakes, and then manage the fallout. You'll do some things perfectly, in ways that only you, the parent, can appreciate as being right. You'll learn as you go, and as your children grow. Over the years, there will inevitably be times when you will discover better ways to care for your children. Instead of looking back with regret, concentrate on the present with increased confidence, and look to the future with hope.

For Friends and Extended Family: Their Family

The fact that you're reading this book suggests that you feel empathy for the children and concern for their welfare. The information and advice offered here will give you a better sense of what the family is going through. Understanding what each parent may be feeling allows you to offer genuine comfort and solid advice when they are upset, confused, frustrated, angry, or exhausted. Even when there are no adequate words to offer at a particularly bad moment, your insight will make you a more sympathetic sounding board. And, you can reinforce the work of the parents in your own interactions with the children. In the jargon of support groups, this book will help you to be an effective support person.

Being a support person carries its own hardships. Unlike the parents, you can't make the decisions or take charge of the problems. Since you are "just" a friend or relative, you may be struggling to figure out how much you can do or say without imposing on the family's personal space. On the one hand, you fear pulling back from the situation too soon and leaving the children to suffer unnecessarily when there may be more that you could do. Years from now, you don't want to be sorry that you didn't do more. On the other hand, you don't want to push too far and cause harm. You need to recognize when it's time to let go.

Here is some simple advice. Read this book. Premise all of your words and actions on your belief that the parents don't intend their children any harm, and that you respect their right to choose how they handle *their* crisis. Make it clear to the parents that they are in control of how much you help, and that you will respect their wishes. Gain their trust that you will never say anything to the children without their permission and knowledge.

Let them know that you care about the family and would like to share your concerns about their children. If they insist that the kids are fine, and you feel otherwise, there are a number of approaches you can take. The first is to wait a while. Stay alert for windows of opportunity to mention your concerns. Unless the parents ask you not to bring up the topic, try again and again. Offer them this book or other reading material. Many parents are more receptive to new or challenging ideas in the privacy of their home, when they are alone or with their spouse. After a few days or weeks, be sure to ask them what they think of the book. Some parents won't remember getting it because they were still in shock, or they put it away for later and then got distracted and forgot about it. Try to sound encouraging with your reminder without sounding pushy. For example, "I read the book myself and found the stories inspiring for handling our day-to-day ups and downs."

When the chemistry simply isn't good for sharing your thoughts about the children, for whatever reason, try getting the information to the parents through one of their close friends, clergy, or someone else they trust and respect. It is appropriate to let the patient's oncologist or nurse know that you are worried, and to show them this book. These professionals can pursue your concern in the context of the patient's overall care.

If the parents still prefer to keep the children in the dark about the illness, step back, be understanding, and help in whatever way the family requests. You are not abandoning the children; you are showing them that you respect their family.

When treatments are over, continue to be supportive and nonjudgmental. If the children appear to be having trouble, you can once again try to steer the parents toward information or individuals who can help. But the boundaries of what you could do for them during treatments still apply. Ultimately, it is their life and they have a right to live it their way. Recognizing and respecting your limits in their lives is one of the ultimate expressions of love.

PROLOGUE

Our Story

"First comes love, then comes marriage, then comes baby in the baby carriage." The limerick to which I jumped rope as a child became a fulfilled prophecy with the birth of Rebecca five years after I married Ted. Two years later, we were blessed with Jessica, and, another two years later, William. Our family now complete, Ted and I knew we'd live "happily ever after."

For me, "happily" meant figuring how to balance the demands of my solo practice of internal medicine, Ted's career as a university professor, and the ever-changing needs of our growing children. In the fall of 1990, soon after turning thirty-six, I was still climbing the steep slope of the learning curve when I was hospitalized with excruciating leg and back pain.

The day following emergency exploratory surgery, my doctor came into my hospital room and explained that I had non-Hodgkin's lymphoma, a cancer of the lymph system. I knew what I had to do because, unlike the average patient, my training as a physician had taught me the ropes of obtaining excellent medical care. Unfortunately for me, I also knew what lay ahead. I'd been privy to my patients' cancer experiences. Treatment for lymphoma is an ordeal that changes a person's life forever after. When the person survives, that is.

With graphic images flooding my mind—baldness, vomiting, radiation tatoos, hospitalizations, surgeries, and dying—my first thought was, *"Oh, no. My children, my children."* I was overwhelmed by the thought of them, not yet two, four, and six years old, suffering emotional damage because of my illness. The idea of them possibly losing me was unbearable.

I wasn't the only one traumatized. While I was lying in the cocoon of my hospital bed, focused on my leg pain and fear of dying, Ted was dealing with the normal world and our very normal children. Ted told me how he drove home alone from the hospital after learning of my diagnosis. He was dazed, afraid, and worried when he was greeted at the door by our three little ones. While his heart told him to devote himself totally to me, his intellect commanded him to tend to the children. Rebecca, Jessica, and William needed him too, although in different ways.

Instinctively, Ted swore to himself two things: First, not to lie to our children. My illness was going to be a long-term affair and he needed their trust. And second, to try not to think about me when he was with the children because they needed his undivided attention. Our shocking news was their crisis, too, as life-altering for them as it was for us. Ted sat down and told them the news, gently and simply.

Life with cancer and treatments was not easy. During the months following my diagnosis, routine matters such as changing diapers, doing laundry, providing daily meals, and fulfilling car pool obligations were accomplished only with the help of friends and family. Special occasions, such as school plays and ballet recitals, drained physical energies when I could attend, and emotional reserves when I couldn't. My own anxiety, fear, and fatigue made the normally challenging ups and downs of rearing children even more so.

How could I possibly be a good parent under such oppressive circumstances? The answer began with accepting that I did not have a choice about cancer entering our lives, and deciding to see my illness as a strengthening force for my family. With these first two steps, something magical happened: in certain ways, parenting became easier than it had ever been before. By making me painfully aware of the fragility of life, cancer clarified my parental duties, restructured my priorities, and taught me how to live in the present, in my children's current moment.

The dreaded cancer invading my body inspired me to embrace the parenting I *could* do each day and generated an urgency to create a wealth of life-enhancing examples and experiences. The way I saw it, the lessons and loving memories the children stored of our todays were my insurance that they would grow strong and feel my love through all their tomorrows, no matter what happened to me.

Nurturing Rebecca, Jessica, and William nourished me. Daily hugs and developmental milestones became celebrations I felt a thousand times over. Unwanted circumstances became opportunities to teach them values and demonstrate unconditional love. Spilled milk, rained-out picnics, and skinned knees became welcomed occasions to help them learn how to deal with stress, change, and loss.

This magic did not happen overnight. A few weeks after my diagnosis, my children's delight in the flickering Chanukah candles triggered great anxiety in me: "They might kindle lights without me next year." Far from savoring the seasonal love and joy, I was swallowed by fear, anxiety, and grief.

Later that evening, when Rebecca got out of bed for the twenty-millionth time and came into our room, I saw her beautiful, doll-like silhouette as an annoying intrusion, preventing me from getting the attention and rest that *I* needed. Half-scolding, half-pleading, I led her back to bed, tucked her in one more time, and said between clenched teeth, "Darling, if you can't fall asleep, look at your picture books for awhile. But *please* stay in bed. I need time with Daddy." What I was thinking—in fact, screaming—in my head was, "Don't you know that I have cancer? I feel miserable. I'm upset. Why can't you just take care of yourself and leave me alone?"

Although my belief in the potential positive power of my illness was born within weeks of my diagnosis, the magic was a gradual process that depended on me, and not the swish of a fairy's wand. I had to learn how to live with my disease and better understand how to raise my children. Even after I felt empowered to parent well, not *every* hug or smile was appreciated, let alone celebrated. There were times when my own emotions or physical sickness kept me from saying or doing what I knew to be the right thing. Through it all, the best I could do was the best I could do.

Ted and I spent considerable time discussing how to educate our children about my health situation in particular and about cancer in general without making them afraid. One of our biggest problems was determining the needs of each child. My two-year-old, William, had very few special needs and no concept of the significance of my illness. For him, cancer simply meant Mommy was home more, a good thing, all things considered. Four-year-old Jessica needed extra physical contact and had a real sense that this illness was important. But she lacked the maturity to really understand what was going on, or why it was out of the ordinary. Rebecca also needed extra physical closeness, but her need for emotional reassurance was even greater. At six, she grasped from the beginning the seriousness of my leg pain and my not being able to work. She was just beginning to understand and worry about the concept of death, and gradually came to understand the life-threatening nature of my illness. The usual adventures of kindergarten became more emotional and complicated in the setting of cancer.

While teaching my children about cancer, I did my own research. During the seven months of chemotherapy, I distracted myself from the discomforts and stresses of the treatments by reading much of the professional and lay literature on cancer and then writing about this common illness. This diversion into writing was self-healing both for what I gained in knowledge, insight, hope, and comfort and for what I could give to others. Too sick to care for patients in my office, I could fulfill my childhood dream of being a physician through books.

The manuscript of *Diagnosis Cancer: Your Guide Through the First Few Months* was completed by the time I finished my chemotherapy and was pronounced in remission. I then passionately threw myself into researching and outlining the manuscript of *After Cancer: A Guide to Your New Life* because it was painfully clear that the cancer experience does not end with the last treatment.

As I struggled through my own recovery, and began to write about life after cancer, one realm of survivorship kept crying out to me—the children. During many hours of private counseling with the hospital social worker, my concerns often revolved around how my children were doing and the tensions I felt as a mom recovering from cancer. Between these sessions, there were innumerable phone calls with two other moms whom I met through a support group, who also were recovering from treatment. Susan's fourth son was just two weeks old when she was diagnosed. Sharon's second child was four months old. Our oldest children were all the same age. Telephone calls were volleys of frustration, fear, anxiety, sadness, and confusion, which played around the same questions: "What is happening to my children?" and "How can I help them?"

Trips to the library and bookstores turned up plenty of books for teaching a young child about death. That definitely was not what I needed or wanted. There were only a few short illustrated children's books about a parent with cancer who recovers, and fewer adult texts offering any information or advice about caring for my kids. The children's books didn't attempt to go beyond a superficial discussion of what cancer was, the most obvious effects of cancer treatments, and the children's emotions. Unlike these tidy storybook tales with happy endings, my illness was drawn out with a rocky recovery and an uncertain future. My daughters and son needed more than what was offered in the available books.

Even when a parent's course is relatively short and smooth, the children need more detailed medical information and practical advice to help them deal with the stresses, changes, and losses. I toyed with the idea of writing some sort of kid's guide dealing with a parent's cancer—a book for children like Rebecca, Jessica, and William. But I needed a different format from that of my adult books if the message were to be meaningful for youngsters. Unsure how to proceed, I put the idea on the back burner.

A year later my first book, *Diagnosis Cancer*, was published. While awaiting the arrival of a newspaper reporter who was doing a piece on the new book, Rebecca, then seven years old, asked, "Is she going to interview me?" "No, honey, but I can," I responded intuitively. "I'll use your stories and advice to write an article to help other children whose mom or dad has cancer." Although my premise was true, my primary motive was to encourage Becky to talk. She needed to talk.

"Tell me things that are hard for you when I'm sick. . . . Tell me things that help you," I asked in my pretend interviewer voice. Becky freely acknowledged a few of her unpleasant emotions while demonstrating some understanding of several of the coping skills we had developed together during the months of my chemotherapy: her frustration with my hospitalizations and slow recovery from treatment, her fear of cancer that prompted visualization with a lollipop symbol, her sense of uncertainty about the future, the dawn of her realization about the unfairness of life, and her lingering feelings of sadness and anger.

After she read the first draft, I asked as nonchalantly as I could, "What stuff did I leave out?"

She responded, "You know. Being scared about . . . you know." Yeah, I knew: *death and dying.* Two weeks before the interview, eleven months after completing my initial chemotherapy, I was diagnosed with recurrent lymphoma. During the following weeks Rebecca shared feelings about death in general, and my death in particular. She never came straight out and said, "Mommy, I'm scared that you are dying." Over the course of a number of days, during which period she was irritable and cried often, we talked about the changes going on—the cancer, the radiation treatments, the decision to close my medical practice, the need for her grandparents to move in for a while to help out, and so on.

Every night after I tucked her in, as was our usual routine, we'd talk together before I sang her lullaby. I'd ask her to tell me about her day or anything else she wanted to talk about, making sure to address any questions about what was happening to me or her. One night she started asking questions about one of our dogs, Curie.

"How old is Curie? How long does that kind of dog usually live? When do you think Curie will . . . you know . . . ? What's going to happen if she . . . you know . . . ?" She then asked about her grandparents, how old they were, how long old people usually live, what was going to happen after they . . . you know.

"Are you worried about me, Rebecca?" I asked calmly. She

started to cry and confided, "If you die I will miss you so much I won't be able to stand it. I'll never feel happy again."

At that most frightening time for me (facing my first recurrence), I reassured Rebecca that even if her worst fears (and mine) happened, and I died, she would be taken care of, she would continue to receive lots of love from remaining family and friends, and she would feel my love for the rest of her life. Through my own watery eyes, I looked into hers as I explained that my death would make her very sad at first and she would always miss me, but she would get used to it and not feel so bad after awhile. With time, she *would* be able to feel happy again. We hugged for a long, long time.

I concluded the discussion by reminding her, "I'm not dying now, Rebecca. We are expecting the treatments to get me well again. I *promise* to tell you if my doctors ever think the treatments aren't working and I'm getting worse. If ever you want to talk about it, just tell me. OK?" She nodded. "Becky, as long as I'm doing OK, let's think about how to get through the treatments and have the best summer we can." I sang her lullaby and kissed her goodnight.

Having Rebecca face her worst fears, and realize that she would survive them, calmed her terror and mine. I said and did the right thing that night. How did I do it without falling apart? Time-consuming preparation had taken place in the months before that conversation: I'd looked inside myself over and over for what I believed. Ted and I had discussed the possible responses. In my head, I'd experimented with various ways to express my thoughts and feelings and then rehearsed the words that sounded best. And the night of our pivotal chat, I cried after she fell asleep.

As painful as it was for Rebecca and me, facing our worst fears allowed us both to move forward. Rebecca thought that by taming these fears, she was "done" with them. I knew better. No matter what happened, but especially if I developed future problems, she would be facing these fears again. Dealing with them up front helped her cope that week, but it also laid a foundation for getting through future troubles. Ever since that night, it has been easier for us to talk about death. We don't need to very often because we've found honest responses to her questions like "Is Mom going to die from her cancer?" and "What will happen to me if Mom dies?" We don't like the answers, but we *can* live with them. And we do.

Rebecca's response to the first draft prompted me to give a worry cup to my daughter and her storybook counterpart. When a parent has cancer, children wonder and worry about death, on some level, whether they talk about it or not. By dealing with these difficult questions in *Becky and the Worry Cup,* I hope to encourage other families to find answers they can live with.

Over the next four years, I periodically reworked the manuscript, and then asked Becky what she thought. Each time she revealed a little bit more about herself. For example, the first draft addressed her feelings about my appearance by having Becky in the story say, "I always thought Mom looked pretty in her scarves." One year later, I was diagnosed with a second recurrence and decided to commute to California to take part in a clinical trial. Before I left, we took out *Becky and the Worry Cup.* This time, as we read the section about the scarves, she interrupted me and confided, "Mom, that's wrong. I never thought you looked pretty in your scarves. You looked weird and I was embarrassed to be seen with you. Didn't you ever notice that in the stores I wouldn't stand right next to you?"

I had never sensed her discomfort with my appearance, even in retrospect. At the time that we worked on the first draft, she felt obligated to be kind about my appearance, and guilty about her real feelings. Maybe it was her being older, but I believe it was her anger about my second recurrence and my leaving her to get cancer treatment in California that removed the inhibition to revealing her true feelings. If Rebecca felt that way, surely other children had similar feelings.

Almost three years after interviewing Rebecca, I read the latest draft of *Becky and the Worry Cup* to William, now a vivacious six-year-old who had heard none of the earlier versions. He remembered little of the actual events surrounding my chemotherapy or radiation treatments, and seemed eager to learn about my disease. Ironically, a few weeks later a routine checkup revealed progressive disease for which a course of chemotherapy was advised. When I informed my children, William said nothing.

That night, after I'd put him to bed, I began my daily exercise workout in my bedroom. Ten minutes later, William silently crept in my room with an armful of new drawings. He laid them out on my bed, leaving a big paper heart on my pillow. When I caught sight of him, he explained with a smile, "I'm practicing for when your counts are low."

In *Becky and the Worry Cup,* I explain that low blood counts from chemotherapy increase the patient's risk of catching infection. I then offer suggestions for how children can overcome their loneliness and sense of helplessness during the times when they can't be too close to their ill mother or father. However confused or frightened William might have been about the news of my impending chemotherapy, he felt he could do something about it. *Becky and the Worry Cup* shows children that they *can* do something about the monster (cancer) in their home.

Becky and the Worry Cup reaped the benefits of the many ups and downs of my illness. Each climb on the roller coaster of cancer afforded me a new view, or a better view, of what was happening with my children. Each hill was an opportunity to practice what I'd learned, and find new ways to help them with all the little and big problems.

When a Parent Has Cancer, the companion book, began as a short "postscript for parents" to the original children's story. Just as *Becky and the Worry Cup* grew with each passing year that my family dealt with cancer, so did *When a Parent Has Cancer.* Other families can choose to leave some issues unexplored when the parent's illness is brief and uncomplicated. Ignoring the breadth and depth of family problems was not a realistic option for me. Since I didn't get well, and I didn't die, my family was stuck dealing with cancer. I had to continue to learn about what was happening with my kids, and find new and better ways to help them. The long ordeal also meant that I had the luxury of time to find out what worked and what didn't, and to learn from my mistakes. In keeping with my lifelong tendency to look for silver linings, every time I discovered something about parenting that was practical, philosophical, or simply beautiful, I added it to the manuscript.

For most of the time between 1990 and 1998, I had cancer. Unexpectedly and amazingly, the eighth round of treatment I received in the summer of 1998 prompted a remission that has persisted to the present, giving new meaning to my oldest mantra, "There is always hope." Follow-up scans and doctor visits occur less often and have become so humdrum that I recently had to leave notes to myself on my refrigerator and bathroom mirror—"Scans today"—to keep from forgetting and missing my appointment!

During all the years of raising my children while going through cancer treatment, I knew that only time would tell the long-term effect on them. Now, some of the effects are apparent. For instance, one day my son came to me with an innocent heat rash. As I examined the little red spots, he asked, "Could it be cancer?" Although I find it sad that cancer would even cross his young mind, I think it's terrific that he didn't hesitate to show me his concern. Thanks to our well-established bond of trust, I could reassure him easily, and he could (and did) let go of all needless worry. My children may not enjoy the bliss of ignorance or the comfort of denial because they are too well aware of the possibility of developing a serious illness. For me, the reward is that they seem to know how to benefit from early intervention.

It's important to note that my son did not appear at all panicked by the thought that his rash might mean cancer. I suspect it's because my kids appreciate that not all cancer is the same, and that cancer is not a death sentence. Their hopeful perspective was evident when a colleague of mine was diagnosed with lymphoma. In sharp contrast to the reactions of all the adults, my middle daughter blurted out, "Now he can be a survivor, too!"

Gratitude permeates the lives of my children. The night after shopping for a gown for homecoming, my oldest and I were saying our good-nights. Out of the blue, this child who rarely if ever mentions my illness said, "I had so much fun with you today. If you had died, we wouldn't have had that." Her awareness of what might have been lost made her truly appreciate a moment that I suspect most teenage girls take for granted.

Empathy springs forth like running water. In the middle of a high school physics lesson, Becky discreetly left the classroom after another student had dashed out without permission. Although my daughter was not a close friend to this girl, she knew that the girl's mother was in the hospital. Becky followed her into the bathroom, and comforted her, knowing from experience what to say and what not to say. If Becky's teacher hadn't called me, I never would have known about the incident. When I mentioned it that evening to Becky, she shrugged as if I'd complimented her on brushing her teeth before bed, saying, "It wasn't anything special." For her, getting help when you need it, and giving help when you can are just part of the ebb and flow of the world.

When bad things have happened in my children's lives—such as when it rained the day of a swim party, or an unfair call by a referee resulted in a team loss, or a sore throat and fever developed the morning of a long-awaited school dance—I never once heard them ask, "Why me?" It seems that they've learned and accepted that bad things happen and that life is unfair. Freed of punishing self-blame, I've watched them grieve their losses, try to change what can be changed and accept what can't, and move forward with energy and hope.

Of the things that can't be changed, one family challenge has been dealing with my limited energy. Believe me—it's hard on my kids. They have to adjust their schedules and sometimes miss out on activities because I can't participate or drive them. And, they pay the consequences of the mistakes I make when I'm tired. Just ask them about the times I've forgotten to do things I've promised to do, like pick them up after school or mail in a school form before the deadline. Yet, every day they show me how much they love me—limits and all—and they readily forgive all my mistakes. When the tables are turned and one of them screws up, they ask for and expect forgiveness. We all try our best. We all make mistakes. We all move on, together.

I could go on and on about the positive outcomes of my family's experience with cancer: my children's ability to find comic relief in painful situations; their willingness to talk with Ted and me about topics most teens find too uncomfortable; my trusting them completely when dealing with sticky situations; the mutual calm that comes with knowing that we'll deal with whatever lies ahead, cancer-related or not. Living with cancer has been hard, and my future remains uncertain as I deal with some ongoing medical problems and my risk of another recurrence. Cancer, however, has never been the sole focus of our life. Cancer is just one part of our life, a life that knows joy.

By finding a healthy philosophy, I have been able to raise my children through their impressionable years better than I might have had I never been sick. But it's been hard. Between checkups, I have to repress fears about my future and consciously focus my energies on my family and writing. As each checkup draws near, or as medical problems develop, my defenses and skills falter. The rising fear bred by all the "what ifs"—What if the cancer is worse? What if I become debilitated? What if I run out of treatment options?—battles my mantra, "I must deal with this courageously for my children."

Trying not to be afraid of something that terrifies me, I cry, "I wish I were braver." Sounding rather like the cowardly Lion in the story *The Wizard of Oz*, I recall how he found his courage. Soon after Dorothy presented to Oz the broom of the melted wicked Witch, the Lion stepped forward. Trembling before the throne of Oz, the Lion asked for the courage that he had been promised. With a booming voice coming from the empty throne, the invisible Wizard put him off, ordering him to come back the next day. Just then, Dorothy's dog, Toto, accidentally exposed the old man behind the screen. Realizing that the wizard was a charlatan, the Lion thought he'd lost all hope of ever being brave.

The Lion was wrong. The little old man behind the screen may have been a very bad wizard, but he was—as he told the Lion—a very good man. He assured the Lion that even though he couldn't "give" him courage, he still could help him. With words of wisdom, not incantation, he revealed where courage is found: "All you need is confidence in yourself. There is no living thing that is not afraid when it faces danger. True courage is in facing danger when you are afraid, and that kind of courage you have in plenty." Cancer can be like the Wizard, revealing the bravery that lies within each of us. My family can deal with whatever lies ahead. So can other families in which a parent has cancer. All we must do is look inside, and find the courage to face the future honestly, with love and hope.

How to Use *Becky and the Worry Cup*

Whether you are the parent with cancer, the well parent, an uncle or teacher, or another adult who cares about the patient's family, *Becky and the Worry Cup* can help you deal with the children in a healthy way by offering sound information, practical advice, and genuine comfort. In this way children of all ages are encouraged to cope and grow strong through the changes and stresses brought on by their parent's illness. The rest of this chapter is written to the parents because you are the ones most directly involved. The information applies to all adults dealing with the children.

The Child's Age

In most cases, children eight or nine years old can read *Becky and the Worry Cup* independently. Many teenagers may find it informative and comforting, even though it is written at a third-grade reading level. Older children have many similar questions and concerns as their younger counterparts when cancer is a new experience for the family. Children four to eight years old can have an adult, preferably a parent, read it to them. Toddlers benefit when you adapt the stories and philosophies to the level of these little ones.

Read the Book Yourself

I recommend that you read *Becky and the Worry Cup* completely before sharing it with your children. Advance knowledge of the story's contents will allow you to direct your children back to the pages when their questions are addressed in the text. More important, in order to be a source of strength and comfort to your children, you must first absorb the information in *Becky and the Worry Cup* yourself. Think about how the facts, philosophies, and advice relate to the particulars of your family. Take time to experience and process your own emotions. The forewarning, "Physician heal thyself," applies well to anyone wishing to help children. Until you begin to tame your own fears and anxieties, release your anger, feel your grief, and answer your own questions, it's hard to help children in a meaningful way.

Arrange a Healthy Reading Environment

Becky and the Worry Cup can be read one or two chapters at a time, or from cover to cover. When possible, read it with each of your children privately at least once. Choose a time when neither you nor your child are overtired, hungry, or distracted. For young children, make sure to leave enough time after reading to shift the focus away from cancer before they go to sleep. Even if your children are comforted and inspired by the story, they may have a hard time suppressing worries or thoughts about cancer if they are expected to go to sleep right away.

With your older children, on the other hand, bedtime may be the optimal time. That is when they are winding down from their various activities and distractions. Unlike younger children, they usually are able to talk about a difficult topic such as cancer, and then switch to a more neutral topic. Help them leave the seriousness of your illness behind as they prepare to fall asleep by talking to them about school, friends, an upcoming holiday, or other happy situations before saying goodnight.

When children prefer or demand to read *Becky and the Worry Cup* by themselves without you, let them. But I caution against leaving them to interpret the story by themselves. I suggest that you make talking about it afterwards a precondition to sharing the book. If they resist, you can say something like, "Becky gives a lot of information and ideas that can mean different things to different people. I'm going to have a tough time unless I know what you think about the things that happen in *Becky and the Worry Cup.*"

Do your best to minimize interruptions. When your children's defenses are down, you must be physically and emotionally present to support them and help them find the tools to cope and grow. The children may suffer emotionally if, after you encourage them to share their feelings, you run off to pick up a crying baby or answer the phone before you've helped your child begin to tame the exposed fears, anxieties, sadness, or anger.

Take your time reading *Becky and the Worry Cup.* Allow your children to respond to it as it is read. The goal is to connect you with them, and encourage their personal exploration and education about cancer, illness, loss, and so on. Encourage their interruptions—their reactions to or ideas about what is happening in *Becky and the Worry Cup* or their own life. Questions, concerns, fears, and uncomfortable feelings may arise that can be expressed more easily in reference to the story's character, Becky. For example, a child may feel more secure asking a parent, "Did Becky cause her mom's cancer?" than asking, "Did I cause your cancer?" Or, after hearing that Becky doesn't like her mom's bald head, a child might venture a comment about his or her parent's appearance, looking for your response.

Be prepared for the possibility of reopening old wounds, such as feelings associated with the earlier loss of a parent to divorce, the death of a grandparent or other loved one, or a move to a new city far from longtime friends. Point out how your family is different from Becky's, and show them how your family meets some of the same needs as Becky's in different ways. For example, children of single parents don't have a healthy parent at home to offset the fear of abandonment. Show them how they can build strong relationships with nonparent family and friends. Help them see the specialness of parent substitutes ("Becky had one dad, you have two uncle-dad's").

Use the Story Again and Again

Becky and the Worry Cup can be read and reread at any stage of the cancer experience. Different parts of the story may help only after a few readings or after some time has passed, depending upon the circumstances at home. It may take a few readings over time because your children's emotional and intellectual maturity, readiness to deal with certain issues, and individual needs are changing constantly. For example, the reference to operations may slip by unnoticed if you didn't need surgery or if your children were infants when the original surgery was done. Now if surgery is needed, the issue is immediate, and a new reading of the appropriate chapter, or the whole story, may help.

Children's anger, denial, disappointment, magical thinking, frustration, hopelessness, or depression may need attention at any time during or after completion of the parent's treatment. By periodically reading *Becky and the Worry Cup*, you can tap into your children's emotions and problems that are closer to the surface at that moment.

Repeated readings also reinforce information and philosophies that are understood but not fully appreciated. Children have a pervasive tendency to blame themselves for any or all aspects of a parent's illness. Even if this issue is discussed and seemingly resolved at the time of the diagnosis, the child's tendency to feel responsible for problems may arise again and again. After completion of treatments, most patients experience aftereffects such as fatigue, and some have complications such as infections. Your children will need reinforcement of basic information, and ongoing support, in order not to blame themselves for causing the posttreatment problems. Noncancer problems that often accompany or follow serious illness, such as marital or financial strain, also lend themselves to children's self-blame unless you actively intervene.

Use the Book to Help You Know How Your Children Are Doing

Use *Becky and the Worry Cup* as a tool to check how your illness is affecting your children. When they read that Becky feels mad, they have an opening to express similar feelings. When Becky shares specific fears, your children's similar anxieties are validated as normal. If your children are not afraid of some things that bother Becky, their sense of mastery will bolster their self-esteem.

Their reactions to *Becky and the Worry Cup* will give you insight into how they are doing. Pay attention to their body language as well as their verbal responses. Are they interested, bored, or distracted? Do they want to hear more or stop reading? Don't push it if your children appear uncomfortable or ask to stop. Let them maintain control over how much you read. Offer them a way out by saying something like "You might be too tired for a story tonight" or "This cancer stuff might be too new to read about right now." Yet always provide an opening for trying again. "We can read it next week—you might enjoy it then. Becky has some neat ideas for us."

Appendix 1 of *When A Parent Has Cancer* is a summary of the developmental stages of children. You might want to glance through these tables before reading either book. Appendix 2 is a glossary that offers simple-to-understand definitions of words children invariably hear when a parent has cancer. This is a reference for when you are unclear about the meaning of new words or stymied about how to explain a word or concept. Be sure to review Appendix 3, a list of resources for further information and support. Appendix 4 will direct you to further reading.

You might want to give your child a worry cup. Have each of your children pick out one of your mugs or bowls, and designate it his or her official worry cup. This is a tangible way for your children to begin to tame their fears, especially if they feel overwhelmed. You can be as creative or fancy as you want—for example, by decorating the mug with stickers or paint pens.

You brought home this book in a loving attempt to educate, comfort, and inspire your children or the children of someone you love. Use *Becky and the Worry Cup* in whatever way works well.

WHEN A PARENT HAS CANCER

1

Meeting Your Children's Fundamental Needs

Turning Problems into Strengths

Cancer has entered your life. You may be recovering from a biopsy or surgery, in which case you are probably feeling the pain, grief, anxiety, and fear that make time seem to stand still just when you need to be moving quickly to gather information and make vital medical decisions. Or you may be in the middle of treatment or recovering, in remission or facing recurrence. Many of you feel totally overwhelmed by what is happening. Standing in the wings of the cancer drama are your children. Parenting instincts urge you to shield your precious children from the crisis, just as you would cover their eyes if a horrible crime were to unfold in front of them.

The problem with this approach is that there is no way to protect your children from the fact that cancer has entered their lives. Even though the cancerous cells reside in your body alone, **the cancer experience is happening to your entire family**. Your children sense that something major has happened; they are trying to understand what is going on and find ways to cope. But they are just children, lacking the maturity, skills, and experience to deal with your illness. If you exclude them, they may draw inaccurate conclusions or find maladaptive ways of dealing with your illness. **If they are included in the crisis, they can be guided toward accurate, healthy, and hopeful interpretations of the events and learn adaptive coping skills.**

Childhood experiences mold the adults that children become. As a medical doctor, I cared for adults with health-related phobias, unresolved anger, or the inability to trust, all stemming from unpleasant childhood events. I also cared for people who *were* able to overcome illness and loss effectively and enjoy fulfilling lives owing to valuable lessons learned from similar childhood crises. As a mother, I have watched my children grow through the challenge of my illness.

Your illness *will* affect your children. **Whether the overall impact of your illness is positive or negative will be shaped by you—your words, actions, and love.** Recognize the powerful role you play in molding your children. Just as important, **understand that you can only affect, not control, how your children turn out.** Many factors that are beyond your control will also influence your children's reactions and ability to cope. A parent whose child has attention deficit disorder (ADD) or diabetes isn't to blame for the problem. In the same way, don't accept blame for problems that arise due to your illness or your handling of it. Try to do your best, and then gear yourself up to deal with whatever happens down the line.

You can learn from others how to handle common issues that arise when a parent has cancer. Sometimes the best answers are hidden deep inside your heart. Providing healthy responses to your children's questions and difficulties can prevent or minimize problems for them in the short and long term. Fewer problems with them means less stress for you. And in your soul-searching for wise answers for your children, you may discover handles that bring *you* comfort, nourish *your* hope, and encourage *you* to have a positive attitude.

You may feel that the enormous demands of your illness are an insurmountable obstacle to instilling values and beliefs in your children or providing them with adequate love. Your heightened physical and emotional needs may lead you to see parenting as something that will just have to wait until you are healthy again. **Your children can't wait.** This is the only childhood they will ever have, a crucial time of development. Choose to see your illness not as an obstacle but as a powerful platform from which your messages are amplified, helping your children understand and believe you and feel your love in a powerful way.

Rearing children challenges your sense of control over your world. Innumerable parents are made to feel powerless by a baby who won't stop crying, a toddler who won't stay in bed at night, or a teen who refuses to clean his room. Cancer, especially during the tumultuous time of a new diagnosis, disrupts this sense of control even further. Your parenting job does not have to represent one more area of powerlessness. Regain a sense of control over your parenting *and* your life by tackling the issues of raising your children while fighting cancer.

Cancer can direct all your attention and energy inward, toward yourself. Only if you put aside your own worries and feelings when you are with your children can you empathize with *their* concerns and needs and recognize how to help them. This requires that you look at survivorship issues from two different points of view—yours and your children's—and separate your experiences from theirs. Helping them understand their emotions will give you added perspective that will help you with your own feelings. And tending to their needs will help you escape the role of victim.

During this family crisis, you may be highly sensitive to all of your children's moods and behaviors. What other parents may brush off as "just a phase" may seem to you a serious psychological problem. Remember that children grow in fits and spurts. Sometimes they are happy and easy to please, other times they can be moody and impossible to satisfy. Trying on different behaviors and attitudes is a normal part of growing up, and will happen whether or not a parent is ill. You will be facing the challenge of distinguishing children's difficulties that would have occurred anyway from those that are related directly to the strain of cancer. You will sort out when to intervene and when to leave the problem to resolve itself. In learning how to do this you will become well equipped to deal with the normal ups and downs of raising kids.

Establishing Open Communication

If I were given two minutes to advise you how to deal with your children, I would urge you to **establish and maintain open lines of communication** about your illness and all the associated fallout. Ongoing dialogue between you and your children prevents and minimizes misinformation, misunderstandings, and unresolved painful emotions. Good communication with you is your children's best defense against inaccurate or frightening stories about cancer or your condition. As long as communication stays open both ways, your children will feel your strength feeding theirs. I'm reminded of a favorite greeting card that says on the cover "It's you and me against the world . . ." and on the inside "Do you think they have a chance?"

Tell your children that talking about your illness or their worries will always be a top priority. Encourage them, with leads such as "A lot of kids feel upset, angry, confused, or worried when their parent goes through cancer treatment. Let us know if you are upset or worried, or if you have questions, OK?" Friends and family may criticize, saying that you are planting the idea in your children's heads that they should be upset. Or these well-meaning adults may imply that you are making it easy for your children to manipulate you whenever they want attention, or want to distract you from another matter, such as when they've done something wrong. Unless your cancer is easily curable, the changes accompanying your illness will inevitably cause your children some distress. Closing the doors of communication carries risks of much greater problems than those associated with your children getting too much attention. Unless they are experiencing the same situation, no well-intentioned outsider can possibly weigh all the factors you have to consider. Rely on your own good judgment.

Teaching your children the art of effective communication is a vital part of educating offspring. It's not as if the only reason you have to bother with it is your illness. The investment you make now in this skill will reap immeasurable benefits for your whole family because the bonds of family trust and open communication will carry over to weathering other difficult or emotionally charged issues such as puberty and sex, drugs, and interpersonal relationships. Happy, healthy families communicate well; your illness can provide the motivation to build strong ties with your children. In the future, you may find facing difficulties easier than other families that were not challenged as you are today.

Keeping the lines of communication open will be an ongoing, active process as your medical situation evolves, social circumstances change, emotional and spiritual issues surface, and as your children's intellectual capacity expands and their understanding and needs mature. You will end up saying the same things to your children in a dozen different ways. Children often need clarification or reassurance at their most vulnerable times, such as when they are tired and getting ready for sleep, not to mention the inconvenient times for you—when you are late for work or about to leave for a sorely needed evening on the town. Sometimes the best response may be to attend to your children immediately and to leave late for work or cancel your night out. Usually you can answer their concerns briefly, with a promise for a complete answer at a more convenient time. No matter what, keep the lines of discussion open. **Good communication is the only way for you to recognize and respond to your children's needs in a healthy way.**

Understanding Children's Fundamental Needs

Children have three fundamental needs that, if met, enable them to adapt to their world no matter what is happening. Understanding these needs will help you say and do the right thing even when you are tired, upset, sick, or confused yourself. These needs are:

1. continuous satisfaction of their basic physical and emotional needs
2. an understanding *on their level* of what is happening
3. reassurance that they will be cared for no matter what happens to you

Taking care of children's basic physical and emotional needs is a never-ending full-time job in every home. No matter what is happening with the ill parent, your children must have regular meals, clean diapers, transportation to school or an important party, words to comfort them when they scrape their knee or have their feelings hurt, and someone to whom they can say goodnight. Your job is to make sure these needs are met, not necessarily to meet them yourself.

Many of children's needs are less immediate, such as their need to learn how to groom themselves or throw a baseball, their need to understand your family's religious beliefs, or their need to feel close to you. These can be met over the course of weeks or months. Good parenting never implies instant gratification of all your children's needs.

Meeting physical and emotional needs alone is not enough. Your children must understand what is going on in their world: why their daddy is in the hospital, why their mommy is losing her hair, and why their mommy and daddy are crying or yelling. Your children are exploring their environment all the time. By figuring out how things work, they make their world a bit more predictable and controllable. Knowing what's going on helps them function and avoid unpleasantness. Since the day they were born, you have been your children's primary source of protection and information as they've ventured farther and farther into new territory. They have looked to you to keep them from touching a hot stove and to explain how to know when stoves are safe to touch. Now, with your cancer diagnosis, they are looking to you to protect them from getting cancer themselves and from doing something that might make you more sick. They look to you to show them how to tame the fears and anxieties that accompany living with a sick parent.

Throughout your illness and recovery, your children need reassurance that they will be cared for no matter what happens to you, and that it is OK for them to be worried about themselves. Children deal better with routine stresses and crises when they are confident that they will always have a home, food, shelter, and love. Assure your children that their birthday celebrations, baseball practices, and school parties always will be important, and you will make every effort to make them happen. However, when circumstances make this difficult or impossible, help them deal with their loss. Encourage your children to continue to spend time with their friends, or save for that special toy they have been eyeing with anticipation. The fact that some of your children's wishes and goals may have to be postponed or abandoned should not discourage them from continuing to make new ones. **Your children's needs and dreams are just as important now as they were before you were diagnosed.**

Telling the Truth

Establishing and maintaining open communication allows you to meet your children's three fundamental needs if, and only if, you are always honest. This is one area of dealing with children where I believe that there is a right way and a wrong way: you must, without exception, **tell the truth in order to establish and maintain a bond of trust.** Your sons and daughters need to be able to believe you, their parents, in order to grow up into adults who, in turn, can trust others. With the added stress and uncertainty of your illness, being unfailingly honest gives your children assurance in a sea of uncertainty.

When my husband, Ted, drove home after learning that I had cancer, he didn't know many facts yet: the days of tests and consultations had only just begun. In addition, he was dealing with his own strong emotions. Earlier in the day, I had looked so awful to him that he had asked my doctors if I was going to make it through the weekend. He was overwhelmed with the fear of losing me and raising our three young children by himself.

Our children were at home, waiting for some explanations. Ted drove into our garage and then, feeling like a student unprepared for an exam, walked into our home and sat down with Rebecca, Jessica, and William. The simple truth provided the guidance he craved. He said, "Mom has cancer. She's going to be sick for a long time."

"Is Mommy going to be all right?" they asked.

"We don't know yet. We hope so. The doctors are going to do everything possible to get her better."

The truth, even when it is painful, can be a powerful healer. Ted would have preferred trying to convince them and himself, "Yes, she's going to be just fine, without a doubt!" The fact of the matter was that he didn't know. Instead of dwelling on the uncertainty, he focused on two hopeful facts: everything possible was being done to get me better, and they had good reason to hope that I would be fine.

Honesty helps your children understand and deal with what is happening. **When the facts are couched in love and hopefulness, you can guide your children toward a life-enhancing perception of reality. In the end, that's what parenting is all about**.

Many parents have been through serious problems before the cancer diagnosis and found that shielding their kids from the ordeal worked best for them. Naturally they are tempted to grit their teeth through the cancer treatments and then put the whole ordeal behind them. **Let me explain why cancer is different** by telling the story of John, a young dad who looked and felt fine after an out-patient biopsy yielded his cancer diagnosis.

John wanted to handle the new illness in the same way as his vasectomy the previous year—privately, without telling his children or friends. Before his vasectomy, he believed that the strain of acting normal over the subsequent few weeks, when he would be physically and emotionally uncomfortable, would be a fair trade-off to protect his children from a situation he felt should be kept from them. His medical course after the vasectomy turned out to be rockier than he had expected. During his recuperation at home, tensions ran high but the entire episode faded into the background within a month. John's family, no worse for wear, returned to normal, leaving John with a confirmed sense of having taken the best approach for his family.

John's biopsy for cancer was more serious, even though he felt better than he did after his vasectomy. Since his cancer was not easily curable with a simple procedure, the magnitude and long-term effects of his cancer experience were going to make it hard to maintain secrecy and continue to act as if the illness were inconsequential. He wasn't going to be able to control all the anticipated rough times of treatment or the unexpected problems and scares during his recovery. Since he wasn't superhuman, having cancer was going to affect how he looked, felt, and acted enough so that his children would notice that he was different.

Not surprisingly, John's family got into trouble. Within a week, the tension was palpable at home. Unlike the temporary situation after the vasectomy, his wife became increasingly distraught and his children unruly. Their explanations for John's symptoms became increasingly convoluted and stretched the limits of believability. As it became harder and harder to keep their secret, their home became a battleground instead of a refuge of comfort and healing. Fortunately, one of his nurses detected the strain between John and his wife, and intervened. After a few weeks of professional counseling, they were able to repair the damage and move forward as a family unit, with honesty, through the rest of his treatments and recovery.

If your children catch you in a deception about your illness, from then on they will wonder, "Is Mommy keeping something from me?" If they are unsure if you are being truthful, they won't be as comforted by your words and actions in the hard times. Your kids will waste emotional energy worrying about things that could have more easily been put to rest.

Even when drastic changes occur—baldness, vomiting, travel to another city—many loving, intelligent, responsible parents find themselves trying to hide the truth. These mothers and fathers treasure their children more than life itself and want to do what is best for them. So they throw a blanket over the elephant in their living room and spend enormous energy distracting everyone from the huge blanket-covered lump, and responding to questions with answers that appear less frightening than the horrible truth.

Unfortunately, when you have cancer, there *is* an elephant in your living room. Fabricated explanations leave room for doubt in your children's minds and make it impossible for your family to deal with the real problems together. If you hold your children's hands and look at the elephant together, although you'll all feel scared, you'll also feel the strength and comfort that come with facing the danger together. You'll be able to combine your resources to help each other prepare for your life during and after treatment, and live it as fully as possible. In addition, logical, truthful explanations make it easier to become accustomed to the unpleasant things. In contrast, ongoing doubts and anxieties make it harder to normalize the changes.

Telling the truth about serious illness is a complicated task, tapping into how you perceive your own situation and what attitude you have adopted in order *for you* to cope. Facts, uncertainties, hopes, beliefs, and philosophies all help to determine what you "know" about this disease and therefore what you tell your children. To help them, you must first learn about cancer. Misinformation and myths about cancer abound, such as "Cancer is a death sentence," "The treatment is worse than the disease," and "People cause themselves to get cancer by dealing poorly with stress." By learning the facts, and finding life-enhancing philosophies, you can provide your children with a sound education about survivorship that will help them, and help you.

Telling the truth is also complicated by the challenge of figuring out how much to say. Children are just children. While truthful answers are essential, it is not always necessary or prudent to offer long, technical responses filled with all the maybe's that accompany cancer and its treatment. In other words, a child who asks, "What time is it?" is not asking how to make a watch. The specifics of what to say to your children are discussed in the following chapters.

When Advice from Others Seems Intrusive or Annoying

Some of you will read this section about open communication and think, "That may be great for other families, but not for mine." And you may find the unsolicited advice of family, friends, and the health-care team intrusive and annoying.

It is important to understand what may be motivating your friends and extended family to persist in mentioning their concerns about your children despite your requests that they stay out of it. Sometimes they push because they are upset that you're sick, and feel powerless against the disease. Out of love and a wish to make things better, they grab on to matters about which they feel knowledgeable. They wouldn't dare suggest what kind of chemotherapy you should puruse, but they may implore you to eat well, pray hard, or work on a healing, positive attitude. Your children's welfare provides another target on which they can focus to regain *their* sense of control over a bad situation that is causing *them* distress.

Try to understand that their heightened emotions may cause them to lose sight of their bounds, without their realizing it. Or they overstep their bounds because it's hard to stop themselves. For example, before your diagnosis, they kept quiet when they thought that you let your children watch too much TV, yet now they are openly critical. There is a good chance that they come to your home with every good intention of respecting your decision not to discuss your illness with your children. As they see you uncomfortable from treatments, or your spouse overstressed, they feel upset and their resolve goes out the window. All you see and hear is their disapproval. You may not see their pain while they're with you; you don't hear their self-condemnation afterwards for having minded your business.

In addition, your friends and family may be tired and overstressed from trying to juggle the demands of their home and work with their desire to help your family. Their intentions are good, but they are trying to do too much. Overextending themselves can create problems in their homes. Put that together with fatigue and inceased emotions, and they may intervene in your affairs against their better judgment. This ripple effect of the illness beyond the nuclear family is brushed aside by family and friends who remark, "I'm not the one with cancer, so I'm not the one under stress."

It is also possible that current tensions are a flare-up of past problems in a new setting. Grudges, jealousies, misgivings, insecurities, and hurt feelings don't dissolve with the discovery of cancer cells. Your sister or brother may offer loving words that to you sound like "another attempt to control my life." Just when people need to be pulling together, families may play out old power struggles without realizing that is what's going on. Consider the following points when people seem to be pressing you to discuss their concerns about your children against your will:

- It is the experience of oncology social workers that children know something serious is going on even when nobody says anything to them.
- Research on children with cancer has shown that they do better when adults talk openly with them than when they are surrounded by a conspiracy of silence. The same openness about your illness can help your children to adapt.
- You may feel overwhelmed by your medical problems. If this is the case, ask people to wait until things settle down before they offer comments about your children.
- It is easy to feel falsely reassured by children who aren't demanding attention, or who respond to your queries about their well-being with a pat answer, "I'm fine." Quiet children are easy children, which is a relief to you when feeling pulled in a million directions. But their silence may be a cry for help.
- In a crisis or chronically demanding situation, it is easy for parents to miss clues that signal trouble, such as their child not sleeping as well as usual, having trouble concentrating, crying easily, eating more or less than usual, or not performing as well at school.
- You may feel that family affairs are private, or that there is something shameful about discussing your children's emotional well-being. Think about other times when you share personal information with friends, family, or professionals for a greater good. If serious illness is new territory for your family, sharing with trusted people can make the journey safer and easier.
- If you find that the temporary shuffling of roles threatens your sense of self, remember that *you* are the decision-maker and problem-solver. These people are just advising; they cannot make decisions for you.

It is also possible that you want people to leave you alone because, although you agree with them, you just can't do what they suggest. You want to talk openly with your children, but every time you try, you get upset or muddled, and you stop before you even begin. It's like the person who loves someone but, for whatever reason, just can't seem to say the words. There is an easy solution: find out if there is a local program for children of parents with cancer, or an individual such as an oncology social worker who can meet with your children once or twice. These professionals can help you establish and maintain open communication.

See this approach as delegating responsibly, not abandoning your parental obligations. A father who faints at the sight of blood may want to comfort his daughter as her laceration is stitched up but is acting responsibly by asking the nurse to do the hand-holding. **One of the ultimate expressions of love is to recognize your limits and put your children's welfare first.**

Perfect Parenting Is an Illusion

This book will suggest ideal ways to help your children through your illness, as determined by the many factors that make you, your children, and your family unique. But beware: perfect parenting is an illusion. There is no *one* right way or perfect way to parent. Raising children is an ongoing education in recognizing and managing problems. All parents learn "on the job," and constantly adjust how they handle situations that arise again and again.

The sense of urgency created by your life-threatening illness may cause you to feel that there's no room for mistakes. This attitude puts undue and unnecessary pressure on you and your children. Life is about making mistakes, surviving them, and learning from them. Children grow from the things you do wrong or half-wrong, as well as those you handle with style. The best you can do is to meet your children's needs in an environment of truth, unconditional love, and trust, no matter what the circumstances. **The greatest gift you can give your children is not protection from change, loss, pain, or stress, but the confidence and tools to cope and grow with all that life has to offer them.**

SUMMARY

- The cancer experience happens to your whole family.
- Establish and maintain open communication. Doctors, nurses, social workers, family, and friends can help you do this.
- Ensure that your children's three fundamental needs are met. These needs are:
  - satisfaction of their physical and emotional needs
  - an understanding on their level of what is happening
  - reassurance that they will be cared for no matter what happens
- Always tell the truth, couched in love and hopefulness.
- Children know when something serious is going on, even when nobody says anything to them.
- Obtain sound information about your illness and its treatment.
- There is no one right way to parent. Don't try to be perfect.

2

Caring for Your Children Through the Crisis of a New Diagnosis

Breaking the News

Your first duty to your children is the formidable one of telling them the news. There is no easy way to tell your children that you have cancer. The task is easier, though, if **your mission** is kept in mind—**to help them deal with their world**. In addition, with everything you say, keep in mind the age of each child, and each child's fund of knowledge and past experiences. Every child is unique, and deserves specialized attention. Of course, you can talk with them as a group when sharing the diagnosis or other news. Afterwards, you may need to spend time privately with one or more of your children, tending to their individual questions and needs.

Timing is important. In most cases, plan on telling your children something as soon as changes occur in the home that affect them. If you are in the hospital, they need to be told something today. If your cancer was diagnosed at a routine checkup, you may be able to wait until you know a little more. In this case, fill them in as soon as you know what you are going to do.

There are three main goals associated with the process of breaking the news:

1. to provide enough facts to allay their immediate fears
2. to reassure them that they will be kept informed and be well taken care of
3. to prepare them for what's coming next

Surround your words with verbal and physical evidence of your love. With everything you say and do, encourage them to keep the lines of communication open with you.

Find a quiet place where you can hug your children and answer their concerns without distractions or interruptions. Tell them the truth and keep it simple: "I just found out that I have an illness called cancer. The type of cancer I have is called [breast cancer, adenocarcinoma, sarcoma, or whatever]." Even though you may feel uncomfortable saying the word "cancer," using it helps to make "cancer" just another ordinary word, and not forbidden, scary, or emotionally charged. Naming something that is confusing or terrifying, and calling it by its name, are two powerful ways to gain a sense of control.

Avoiding the word "cancer" communicates to your children that the problem is too awful even to say aloud. After all, everything else in their world has a name. In addition, if you refrain from saying "cancer," you risk your children hearing it first from someone else in a context that is not hopeful or helpful. As discussed in Chapter 1, maintaining lies and secrets drains enormous energy as you try to protect your forbidden word, and worry about what would happen if your lie-by-omission were exposed.

The word "cancer" usually does not carry the negative connotations for your children that it may for you. Teaching your children that cancer is a disease like other ailments gives them a foothold against prevailing myth and misinformation, and helps to establish a way of talking about cancer that makes it easier to integrate the reality of your disease into the life of your family.

Using the specific name of your disease helps create some distance from people with other types of cancer, some of whom may have had bad experiences. Your children will have an easier time separating your illness from that of Aunt Susie if you can say, for example, "She had breast cancer, and I have lymphoma."

If you have trouble enunciating the type of cancer you have, you can use your speech difficulty as a symbol for the newness of the situation and for your discomfort. Take advantage of the metaphor by assuring your children that just as you will learn how to pronounce the name, you will learn how to deal with your illness. As you and your children master the pronunciation, you will have a concrete reminder that you are making progress. For example, "We're learning a lot about dealing with multiple myeloma. Remember, kids, when we couldn't even pronounce it?"

Depending upon your family's style, you may use your initial trouble with the name as an opportunity for comic relief: although the cancer is serious, the name of your cancer may be a tongue twister that sounds amusing. One father, Michael, had success adding a splash of levity after being diagnosed with adenocarcinoma of the colon. He told his children: "I can't remember exactly what they called this cancer, I think it was something like 'I dunno, cars and oh my!' of the colon." Certain that his children knew he was teasing, from then on he was able to make reference to "the cars in my colon" or "the cars" as a dependable way to diffuse tension about his illness. Laughing about something that is scary, just like naming it, helps your children begin to tame their fears.

If your cancer was discovered at a routine checkup, your children may have a hard time understanding that you have an illness. Confirm for them that you are not sick in the "feel bad" sense of the word, but that something is wrong in your body that was found *before* it had a chance to make you look or feel ill.

If you've known about your diagnosis for awhile but only now realize the necessity of sharing the truth, you can repair the breach and establish trust. First, review the facts: "You know I've been sick for a while. I told you I had an infection. I didn't tell you everything about my sickness, because I thought I shouldn't, and I didn't know how to tell you. I've learned that I must always tell you the complete truth, even if it's sad or scary. I have an illness called cancer. . . ." During future conversations about your health, your children will benefit from a reminder of your vow to be truthful. You might preface new information with something like, "When I was first sick, I messed up and didn't tell you the exact truth. I'm keeping my promise to always be honest." Reminding them too often is less risky than depending on a single past promise to allay lingering insecurities about your truthfulness.

Who should break the news of your diagnosis? In general, it is best for you or your spouse to tell your children the news about the cancer diagnosis, as well as about setbacks, complications, or recurrent cancer, provided you can handle it. Your crying or showing emotions is appropriate and acceptable. You are sharing with them that this is a difficult time for you and teaching them that it is OK to show emotions. Be direct, so your children do not misinterpret what they are seeing and hearing: "Kids, I'm crying because I'm sad that I'm sick. I'm nervous about some of the changes. Even though I'm crying right now, I know that we *can* deal with this, and we'll help you deal with it."

If you are uncontrollably upset, your children will be frightened and feel insecure no matter how perfect your choice of words. In this case, have someone else talk with them, preferably a trusted adult who has an intimate relationship with the children—a grandparent, teacher, or adult friend. There may be other reasons why you are unable to break the news yourself. You may be in the hospital, too sick to deal with the children. Or you may be traveling around the country getting second and third opinions when you first realize that you need to tell your children.

Never lose sight of **your number one priority—getting the best medical treatment possible to get you well**. If someone else has to take care of your children's needs, even breaking the news, while you take care of yourself, so be it. Try to stay as involved as possible with your kids, but don't sacrifice your health. It's better to screw up dealing with your children in the short run, and to take care of them later, than to put all your energy into meeting their needs and to delay (or not investigate) life-saving treatments.

Teaching Children That Cancer Is Not Contagious

After you break the news of your cancer, one of your first jobs is to make sure your children understand that cancer is not contagious. Be direct: "Cancer is not a catchy illness. It is impossible for you to give it to me or get it from me. Nobody in the world has ever caught cancer from someone else." Liken cancer to other noninfectious medical problems with which your children are familiar—their grandmother's arthritis, their friend's broken leg, or your migraine headaches. I continue to reinforce this important point by quizzing my children periodically when a situation lends itself to this purpose. For instance, one of Rebecca's classmates came to our home to play, and saw my books on cancer lying on the coffee table. When she mentioned that her aunt had cancer, I asked Rebecca to tell her friend if she could catch cancer. Rebecca was the expert here, and said definitively, "No!" When Ted had a cold, I told the kids not to kiss him on the mouth, because they could catch his virus. Then I asked, "Can you catch cancer?" They chimed, "No!" Even when my kids didn't believe it completely, repeating the point helped to reinforce the concept that they were safe from my cancer.

You will need to reiterate the fact that cancer is not contagious throughout your course of treatment and recovery. This seemingly simple task may be complicated if you are susceptible to infection following treatments such as high-dose chemotherapy, and need to keep a distance from your children *for your safety*. Your children may too easily blur the distinction between your risk of catching an infection from them, and their risk of catching cancer from you. Be as clear as possible, stating the facts regularly and in as many different ways as possible, such as through songs or drawings. Role-playing with them or with dolls are two other effective teaching tools you can try.

You can't afford for your children to misunderstand this fact. If they are worried about catching your cancer, they will keep their distance, which will be a loss of precious intimacy for both of you. Or they may feel anxious when they are close to you. Whether they stay away or feel uneasy when they are close, they may feel guilty about their behavior and feelings. Even in this day and age, your children may be exposed to adults who express concern about catching your cancer. Indicate to your kids when other people are misinformed, just as you might criticize an adult who doesn't appreciate the importance of using seat belts when driving.

Another factor that may confound your attempts to convince your children of the noncontagious nature of cancer is the fact that some types of cancer run in families (i.e., are hereditary). As our understanding of cancer progresses, we are beginning to learn how genes play a role in the emergence of certain types of cancer. In the meantime, your children are trying to figure out what it means if their dad has the same disease that their grandmother had. Even when no other family member had or has cancer, your children identify with the family. On some level, they may grasp the idea of genetic links, even if by a nonscientific, magical notion of your family bond. Your children, especially the older ones, may be asking or thinking about more than the possibility of being contagious with the question "Can I get this, too?"

If your type of cancer does not run in families, you can tell them, "Some kids worry that they will get cancer because their mom or dad got it, just the way they got their blue eyes from their mom or dad. You don't have to worry that you will get my type of cancer because I have it." Then, back up your reassurance by comparing your type of cancer to some nonhereditary process that your children remember and/or understand: "The cancer just happened to me, the way a broken foot can happen. You don't have to worry that you'll break your foot because I tripped over the dog and broke my foot a few years ago. My cancer is like a broken foot; just because I got it doesn't mean it will happen to you, too."

If your type of cancer does run in families and your children do have an increased risk of developing the same cancer, it is important to recognize that information about the genetics of cancer is changing rapidly. As long as research is funded, you can expect the future to hold better means of diagnosis and treatment. Focus on what you can do, not what you can't control. Worrying now about cancer that may never develop, or may be easily curable by the time it does happen, wastes everyone's emotional energy.

There are ways to address your children's risk of developing cancer without feeding their fears. Most types of hereditary cancer rarely occur before adulthood. And, to kids, twenty sounds old, and forty seems ancient. Once children learn that they are not in any immediate danger, most are able to let go of worry. For example, your children know that they will look and feel older as they age, but aren't threatened by this knowledge because it is a reality that seems far in the future.

Even when your children feel safe from getting your type of cancer, they may be concerned about getting any type of cancer. Your children may come right out and ask if children ever get cancer. Until my own kids asked, I didn't bring it up. However, if yours know of a child with a malignancy, or if they comment on an advertisement for a children's cancer fund, deal with the issue up front. Don't leave their imaginations to fit this piece of information into their world view. Tell them the reassuring facts: most children grow up healthy; most cases of cancer occur in elderly people; relatively few children get cancer; and most children who get cancer are cured.

You can enhance your ability to soothe their concern by punctuating objective statements with your personal mind-set that, in light of the facts, *you* don't worry about them getting cancer. They may persist in asking questions about unlikely or pessimistic scenarios. Try to address their concerns patiently. After awhile, when all else fails, you may help calm them by concluding the conversation with something like, "I'll tell you why I'm not worried—because I'm not!" This is like when you can't convince your children with ten logical reasons why they must go to bed, so you finally persuade them with the notorious conclusion: "Because I said so, that's why!"

Preparing Children for Changes in a Life-Enhancing Way

One of the purposes of providing information is to prepare your children for changes that are coming their way. Children crave predictability. When you know that life is going to change at home because of your cancer treatment, or anything else, warn your children. Reassure them that you will do your best to prepare them, that their needs will be tended to, and that there will always be people available to help them adjust. Tell your children about obvious physical differences you expect, such as hair loss, visible surgical scars, vomiting, and catheters. I told mine, "I'm going to go to the doctor's office every Monday to get medicine in my arm. The nurse will put a Band-Aid over the spot where I get the medicine. Sometimes I will have a black-and-blue mark, but it doesn't hurt much even though it looks yucky." Your children will always see you as their Mom or Dad, no matter what you look like. Demystifying physical changes encourages them to adjust to the specific ones they see, and accept the idea of change and loss in general.

You may not be sure what some of the changes will be. Nobody is expecting you to predict the future. Your job is to help your children cope with their present and not be afraid of their future. Tell them that there will probably be some unexpected changes, too, but that you will keep them informed as much as possible.

Children benefit when the facts are introduced in a framework for interpretation. One day, as I was leaving home to get chemotherapy, the woman who was baby-sitting our children shook her head and said, "Oh, you're going for that nasty chemotherapy." I corrected her, "No, I'm going for the wonderful medicine that is rough to take but can get me better." My words were said as much for my children as for the woman. I wanted to encourage a positive attitude toward the toxic but beneficial therapy.

When you discuss your treatments and procedures, try to focus on their benefits, not the side effects or risks. The story of a woman whom I met in support group comes to mind: Sarina, a young mom with Hodgkin's disease, told her youngsters, "The doctors are going to paint some red lines on my neck and chest that will make me look like a clown. But these lines will help the doctors give just the right amount of radiation to get rid of the cancer. After six weeks, I'll be able to wash off the lines. It's worth looking like a clown if it helps the doctors get me better."

As your appearance changes, your children's friends, teachers, and other associates will probably ask questions or make comments. It might be wise to rehearse with your children what they would say if someone were to ask, "What happened to your mom's hair?" or "Why does your dad have those lines on his neck?" You can help your older children find comfortable responses to questions like, "Why can't your dad drive a car anymore?" or "Why does your mom take naps?"

Putting the disagreeable changes in perspective helps your children adjust to them and offers you an opportunity to instill values. Through the most graphic of examples, look at what you can teach your children about superficial appearance (you are the same person inside, no matter what you look like on the outside), losses (sometimes you have to lose things to get other, more important things), delayed gratification (sometimes you have to wait a long time to reap the benefits of current work and sacrifices), and changes (you can learn to accept and adjust to unwanted changes).

Your dreadful physical losses can be a beautiful way to plant seeds of self-confidence in your children that will help them for the rest of their lives. You can teach through word and example about the essence in each of us. Over the years, my children have witnessed my physical transformations: totally bald and ten pounds overweight from chemo, half bald and ten pounds underweight from radiation, bandaged or limping after surgeries, arms dressed with IV catheters or bruised from failed attempts at getting them in. My standard comment became, "It's worth it if it gets me better. I'm the same person inside even though I'm [bald, fat, whatever]."

On occasions when one of my children's appearance was altered or tarnished, such as when Jessica got her first pair of eyeglasses or when William got a butchered haircut, they commented spontaneously, "I'm the same person inside!" They still felt sad or embarrassed, but the unpleasant feelings were countered with a solid sense of self independent of their appearance.

Having a sense of self that is independent of physical appearance is an asset throughout life. Such people more easily adapt to the usual age-related physical changes as well as to the unexpected ones related to illness or injury. My entire family reaped the benefits of this message when Rebecca, then ten years old, was injured in a car accident. Her face was badly swollen and discolored for weeks. She looked awful and attracted stares wherever she went. Our tested belief in the integrity of the inner self helped us all through this frightening time. **You can't protect your children from the big and little bumps and bruises of life, but through your experiences with cancer, you can help your sons and daughters lay a foundation for surviving them.**

Encouraging your kids to accept your baldness or catheters does not mean that you should rub their noses in it. Highlight ways you can minimize the most disturbing aspects: "The medicine is probably going to make my hair fall out in two to three weeks. If I do lose all my hair, I'll try to have fun wearing pretty scarves or baseball caps while the medicine is working on the cancer." Your children can get used to the bruises on your arms, but still benefit from your wearing long sleeves. Out of sight, out of mind. This way they aren't reminded all the time, and they don't have to deal with others' looks of repulsion or pity.

The value of this was brought home to me during my most recent round of chemotherapy. The drugs given to me in 1990–91 caused complete alopecia (hair loss). I was so much more concerned about being sick and possibly dying that I never shed a tear about losing my hair. During the ten months that I wore scarves, turbans, and hats, the matter of my baldness seemed to be a minor issue with my kids. Last year, when progressive, recurrent lymphoma threw us into the treatment-decision cauldron, my doctor triumphantly presented me with a treatment option that wouldn't make me lose my hair. My response was immediate, and sounded unappreciative: "I don't care if I'm bald! Don't let that play into the decision-making. I just want to get better." Well, of course, the hair-sparing benefit of his choice of medicines did not factor into his decision. Rather, it was a bonus that he felt would make me happy.

To my surprise, having hair throughout the course of

chemotherapy made a big difference. For me, I could more easily slip into social settings and appear normal. My head of hair made it easier to escape the patient role when I wanted to because people would forget, or not realize, that I was in chemo. More important, my children were able to put the cancer out of their minds more easily for moments or hours at a time. That was not the case when I was bald. Then, even though I acted unruffled, and even though my kids got used to the head coverings, my appearance reminded them of my illness every time they looked at me. They spent emotional energy suppressing unpleasant thoughts and feelings that surfaced, even if only on a subconscious level. So, **don't go into contortions trying to protect them from unpleasant changes, but make the effort to shield them from easily hidden daily reminders of your illness.**

Empowering Children

Find ways that your children can help you or help themselves through the crises. Unwanted change is most stressful for them when they feel helpless to do anything about it. Empower your children with realistic choices, remembering at all times that they are children. For example, you might announce, "I will not be able to carpool next week, so Mrs. Smith will drive you to your baseball practice." Although your kids have no choice about who is driving, try to offer them options within the limits of the situation, such as, "Mrs. Smith wants to bring you a snack to eat in the car. Would you like an apple, or crackers and peanut butter?"

In addition to creating choices, encourage their contributions. Teach them how to make their own lunch, and praise their efforts when they do. They can even help you through your times of feeling sick, sad, or discouraged. Invite them to make cards or pictures to cheer you up. Just be careful that you aren't encouraging them to be little adults, or shouldering them with adult responsibilities. It is not a child's place to be a substitute parent, confidant for adult issues, or counselor.

You may find some circumstances, such as losing a limb or having a colostomy, impossible to turn into something positive.

Explain to your children that this is difficult for you, just as it is for them. Reassure them that you will not always feel this way. Remind them repeatedly that the price is worth it: otherwise you wouldn't have done it, and you would have the same surgery again if you had to. Remind your children that, **although you are unable to fix the physical loss or avoid feeling sad about it, you *can choose* to grieve and learn how to adjust.**

Through the months of my own initial chemotherapy, I learned that my kids did better when I was firm about the unavoidable changes they didn't like. For example, when they tried to make me feel guilty about canceling a trip to the movies, in hopes that the edict would be repealed, I learned to say, "I know you don't like it, but there is no choice about going to the movies. Your choice is whether you sulk or find something fun to do at home."

Children are resilient. Their response to news is shaped not so much by what the change is as by how you present it and how well-prepared they feel. If news is presented in a matter-of-fact but reassuring way, your children will feel that they can deal with the changes better than if they sense ambivalence, depression, or guilt feelings on your part.

A cancer diagnosis is as much an emotional as a physical challenge for the patient. While the physicians try to control the cancerous cells, the patient must wrestle with the varied emotional and spiritual issues that arise. When you deal with your emotional "stuff" in a healthy way, you reserve more resources for dealing with the treatments. The same goes for your children. Tending to their emotional and spiritual needs in a timely and constructive manner will better allow you to help them accomplish the job all children face—growing up. That will be discussed in the next sections.

SUMMARY

- If possible, tell your children the news yourself.
- Keep in mind the age and past experiences of each child (see Appendix 1).

- Keep your explanations simple.
- Minimize distractions when you talk seriously with your kids.
- Use the word "cancer."
- Teach them that cancer is not contagious.
- Prepare them for expected changes in a life-enhancing way.
- Help them adjust to changes.
- Shield them from easily hidden daily reminders of your illness.
- Empower your children with realistic choices.
- Encourage your children to contribute to your comfort in an age-appropriate way.
- Your children's response is shaped by what you say *and* by how you say it.

3

Caring for Your Children Beyond the First Few Weeks

Your job of teaching and helping your children adjust to your illness doesn't end after you've discussed the diagnosis and treatment. Life during treatment is usually filled with ups and downs—the proverbial roller coaster of cancer. During the weeks, months, or years of your treatment and recovery, your children are maturing and are able to understand more of what has happened and what is going on. Their physical and emotional needs are changing, as well as their need for information, comfort, and support. Two-way communication about your illness and their needs must remain open at all times, even during those times when you can, or want to, shut off thoughts about cancer.

Learning to Recognize Choices That Are "None of the Above"

Your job is straightforward: to identify your children's physical and emotional needs as they arise and change, and make sure that they are satisfied. However, after cancer, this is often easier said than done.

The situation with one young mother whom I met while on the lecture circuit illustrated this point well. Cindy had undergone intensive chemotherapy, which impaired her immune system, making her vulnerable to infections. She wanted to follow her doctor's advice not to change her son's diapers until her immune system was stronger, but worried that her baby would suffer if deprived of the special nurturing that comes with this intimate, otherwise unappealing task. She felt guilty about not caring for him herself.

Cindy's sadness and anger flared every time he was wet because she was reminded of her lack of control over what was happening. Feeling deprived of the freedom to diaper her baby compounded her sadness over having lost her health and the normal experiences of new motherhood. She considered taking the risk of diapering so as to avoid all the negative feelings and stress she felt when watching other people take care of her son.

Cindy clearly loved her baby and wanted to be the best mom she could. But she had lost sight of how best to serve her child's needs. It had to be pointed out to her that her baby's greatest needs were a clean diaper and a healthy mother. Abstaining from diapering was the right thing to do, and she had no rational reason to feel guilty. Her baby would be fine no matter who changed the diaper. They both would be worse if she developed a preventable infection.

Although she may have avoided her immediate sense of deprivation by diapering her son herself, she would have invited additional anxieties associated with doing something against medical advice. Her anger and sadness about the underlying issues would not have been relieved; they would have been delayed, only to reappear at another time and place. The choice she wanted—to diaper the baby without taking any added risks—wasn't an option.

Cindy was in a variation of the no-win situation that I like to call a "none of the above" dilemma. This is a predicament that requires you to make a decision when each of the options carries a price. The desired answer is, as we used to check on multiple choice tests, "none of the above." In life, unlike school exams, waiting for the impossible is never the correct choice. You risk getting stuck. **Recognizing what you most desire as "none of the above," and then letting go of your desire to pursue the unavailable solution, frees you to deal with the problem**

in a realistic way and move forward. In contrast to the typical no-win situation, in a none-of-the-above dilemma, if you choose the best of the available options, you will win.

For Cindy, the prize for following doctor's orders was maximizing her chance of getting well. With time, she learned how to accept the accompanying unpleasant emotions. She would say to herself, "Although I can't change his diaper, I can play with him afterwards." Her loss was less distressing when she learned to see relinquishing the task of diapering as an act of control over her medical outcome.

There is another trap that can lead you to forego the best choice in a none-of-the-above situation: often masquerading under the guise of "I'm not deciding yet." you don't take any action on a problem, or you continue to do what you've been doing. Not deciding, especially about cancer treatment, may mean losing a window of opportunity.

When I was diagnosed with my first recurrence, I gathered information about all of my treatment options. I didn't like any of them. If I'd kept waiting for my first choice—painless, guaranteed effective treatment—the disease would have progressed and made me a poorer candidate for any treatment. I had to let go of my hope for an easy fix in order to pursue a real option. That's life. The sooner I stopped wanting the solution that was "none of the above," the sooner I was able to regain what power I had over my fate.

The road to cancer survivorship is paved with innumerable decision-making junctures related to treatments, work, and relationships. Many times, every option involves some hardship, loss, and/or pain. **Learn to let go of wishes for impossible options, accept the down sides of your choice, and focus on the positive aspects that made it the best choice.**

Helping Children Through Unpleasant Changes

Cindy's dilemma about diapering was simple compared to the decisions you may face. After all, her baby seemed happy no matter who changed his diaper. When you attempt to delegate some of your responsibilities, your efforts to do the right thing may be complicated by your children's obvious distress. Their reaction is understandable: your driving the car pool and packing their lunches reinforces for them how much they are valued. You are demonstrating a willingness to sacrifice your time and energy for them. In addition, they are comforted by the routine of you driving to baseball practice and your special way of making the sandwiches.

Any disruption of routine can be stressful, particularly if precipitated by a negative event such as illness. Your children's resentment or distress may seem out of proportion to their complaint. But if your children are hysterical because "Dad's peanut butter sandwiches aren't as good as Mom's," for example, the outburst may be a displaced reaction to larger concerns. They may really be saying, "I'm mad that you got cancer," "I'm scared that you're so sick that you can't even make my lunch," or "I feel unimportant when you rest instead of making my sandwich."

In this particular example, your children *need* three things: food for lunch, to feel your love and involvement, *and* for you to do your best to get well. If your children are old enough, explain to them what their needs are: "I understand that you want me to make lunch, but what you need is a lunch *and* a healthy mom. Dad's sandwiches will have to do, because I need to sleep in the mornings until I am better." When possible, invite your children to offer ideas that will make the change easier for them. Ask questions such as "Would you like me to show Dad how you like your sandwiches? Would you like to buy lunches until I can make them again? Would you like to learn how to make your own sandwich?" When you expect the new routine to persist for an extended time, try not to be discouraged if your children don't respond or, worse yet, reject your initial offers. Leave the door open: "I think it will help if I show Dad how to make the sandwiches. Maybe next week you'll want to learn how to do it yourself." Offer them the same or additional choices again and again as the weeks pass.

When I was undergoing chemotherapy, Rebecca invited me to the big kindergarten event in May—Mother's Day Tea. My last treatment was scheduled for April, and we both looked forward to her special day. Unfortunately, we found out just days before the tea that I couldn't go as planned. Because my last treatment had to be delayed, my blood counts were still too low and it wasn't safe for me to be in a small room with a bunch of kids. We were all upset.

What did my daughter really need the day of the Mother's Day Tea? She needed to feel my presence *and* for me to maximize my chance of getting well. Since I couldn't attend like all the other moms, she also needed to express her feelings and be comforted. I made it clear that I wished to go and felt sad about missing it, but that my getting better was more important for everyone. I was firm that my going was not an option. Then we concentrated on the choices still open. The day of the tea, I gave her one of my handkerchiefs "filled with my love" to put in her pocket, and then wished her good luck as she left with her dad and grandfather. They videotaped the program to share with me at home. We made a big deal, in a merry way, about the fact that she was the only one in her class who was allowed to have her dad there. Teasing Ted about attending a function for moms helped us talk about how family love means taking care of each other, doing each other's jobs when necessary, and getting into unusual situations sometimes.

All of our efforts were aimed at making an unpleasant situation less painful in the short run and into something fun to remember: Dad goes to Mother's Day Tea. We acknowledged Rebecca's justified disappointment, sadness, and anger, but did not focus on these negative emotions. Out of a rotten set of circumstances were born memories of togetherness and love. Over the subsequent years, the Mother's Day Tea story became fun family lore. Once my son William was upset because neither Ted nor I could make it to one of his soccer games and a family friend was going to cheer in our absence. Rebecca comforted him with a chuckle, "If you think that's bad . . . when I was in kindergarten, Dad came to my Mother's Day Tea!"

Sorting Out Priorities

The diapering dilemma, peanut butter sandwich fiasco, and Mother's Day Tea are three examples that illustrate how prioritizing even the most mundane tasks can be complicated by a conundrum of emotions. This is not unique to cancer survivorship. Any time there is a major change at home, the stress level can rise as everyone adjusts. When a newborn baby is brought home, the routines change in a noisy, obvious way. So, too, rules and routines are disrupted when one parent is away on a trip, or when parents divorce. Some of the stresses associated with cancer treatment can be more subtle, yet still demand a significant adjustment of your routines, priorities, and roles, often for a long time, maybe from now on. When you understand and respect how the conditions at home have been affected by your illness, you are better able to make wise decisions based on adapted priorities and rules that serve everyone's best interests. People get into trouble when they try to make decisions based on an old set of priorities and rules that can't work under the new circumstances.

Modifying your priorities is not something new to you. Before you had children, when you had an impulse to see a movie, you might just stop what you were doing and go. As a new parent, the same desire to go out to the movies may have stimulated a few moments' thought and then been quashed. Even if you could have found a baby-sitter on such short notice, getting a good night's sleep had become a higher priority than satisfying a whim. You probably wanted to feel rested when your baby awakened the next morning at 5:00 A.M. In other words, the same sudden desire to see a movie triggered a different response because your circumstances were different after having a baby, and therefore your priorities and rules had changed.

In many cases, compromise will allow you to take care of your needs as well as your child's. I remember one time when Rebecca was around nine years old, she asked me to take her to the library to get books needed for a school report. I was exhausted. After trying unsuccessfully to arrange for someone else to take her, I negotiated with my daughter: "Can you wait for me to take a nap first? Or, if we go now, after we get back from the library can you set the table and answer the phone while I rest?" Without apologies, I made it clear that I couldn't do everything, but I would take care of her needs if she helped me. Rebecca was learning about negotiating, sharing responsibilities, and taking care of each other's needs.

How do you figure out which tasks you should do, which you can leave undone, and which you should delegate? The answer is found by asking yourself the following two questions: (1) What does your child really need (as opposed to want)? and (2) Under the circumstances, who is the best person to meet those needs?

At first you may find prioritizing and delegating your responsibilities extremely laborious. Even after you've developed some sense of how things have changed, you may feel stressed because of the unpredictable day-to-day fluctuations in your energy and emotions, and the roller-coaster nature of cancer treatment. Calmness gives way to irritability as your checkup approaches. Treatment is going smoothly when, suddenly, you find yourself in the hospital receiving antibiotics. Each time your specific circumstances fluctuate, you need to readjust.

Creating an "Energy Bank Account"

Another way to help you determine which of your children's needs are best fulfilled by you is to imagine yourself as having an "energy bank account," which holds your emotional and physical energies. Whenever possible, make "deposits" so that you have some reserve for the bad times that often occur. In order to keep your account balanced, you must have some idea of the price of responding to your children's wants and needs. Will taking your children to the movies drain your energies and be a withdrawal from the bank, or replenish you and make a deposit? Will going to the library when you are tired deplete more or less energy than dealing with a child who is rightfully upset if you don't go?

Calculating the cost of your activities may seem like a lot of work at first. Be patient. Your efforts will benefit your entire family because keeping your account in the black will minimize the problems, mistakes, and tensions that occur when you are overexhausted and trying to function. With time, you will start to compute your deposits and withdrawals almost automatically.

Be forewarned—sometimes you'll calculate wrong. For instance, you'll decide to go to the movies, anticipating the warm, fuzzy feeling of family togetherness. Unfortunately, your youngest child may spill his drink all over your older child, as the others complain about their uncomfortable seats in the packed theater, and so on. Those are the times you have to shrug it off and remember that the best you can do is the best you can do. When you overdraw on your energy bank account, pay the penalty and then start making more deposits. This teaches your children how to accept and laugh at such problems, and move on.

When Children's Needs Are Not Met

There will also be times when, despite your best efforts, your children's specific needs will not be met. Your baby may get stuck in a wet diaper all evening, or you may be late picking up your children from school because you couldn't drag yourself out of bed. Efforts should be made to avoid these situations, but do not overplay their negative effects. Babies are not permanently damaged by a diaper rash. Your children won't develop lifelong fear of abandonment because once or twice you picked them up late from school. Again, the best you can do under the circumstances is the best you can do.

Comfort your children while using these episodes of unmet needs to build your children's self-esteem, sense of resiliency, and self-reliance. Teach them what to do and how to deal with their feelings if they are left stranded at school, or if their plans fall through because of you. Applaud their efforts to deal with the challenges. Reinforce their realization that they can handle being hungry, lonely, disappointed, or uncomfortable for short periods of time. Help them to accept your shortcomings and to forgive you. Turn a bad situation into a learning and strengthening experience about hardship, imperfection, and forgiveness. Most important of all, accept your own imperfections. Don't beat yourself up over things that slip through the cracks. It's OK if you screw up. Forgive yourself, and let it go.

Knowing How Much Detail to Tell Children

Your children will have an easier time adjusting if they are kept informed about your medical condition and how you're doing emotionally. I've urged you to tell the truth, always. But, as I explained in Chapter 1, telling the truth does not mean sharing every detail. **Tell them enough, not everything.**

You may find that knowing how much to share with your children is one of the most challenging aspects of raising them through your illness. Inform them about facts and issues that affect their world. Say everything with the reassurance that you will keep them apprised of what they need to know, that you will tell them what they want to know, that you will prepare them as well as possible for changes that affect them, and that their needs will always be met.

As a general guidepost to knowing how much to tell them, think of the world as they perceive it. What do they see? What do they feel? What do they care about? What do they need? What do they want? When trying to figure out how much to say, remember your goal—to provide information and support that will help them deal with *their* world.

Think about the medical issues: Although you may not think twice about a routine blood test, the telltale Band-Aid may signal major disaster to your child. You may be thrilled that your hair is growing back; your child may see this regrowth as a signal that your cancer is growing back. Your continued need for extra rest after completion of treatments may lead your children to feel confused, worried, or angry if all you've told them is that your cancer is gone. As they see it, if you're in remission, you should be back to the way you were before you had cancer. Since you're not, something must be wrong.

The specifics of what you tell each of your children will depend on his or her age, maturity, prior experiences, frame of mind at the time, how much you know about the situation, and how much the child wants to know. Be sure to tell them enough. You need to tell them if you still have cancer after treatment ends, so-called residual disease, or if the cancer has recurred. Tell them when you are going to receive surgery, radiation, chemotherapy, or other medicines, and how the doctors expect the treatments to affect you (e.g., hair loss, fatigue, sun sensitivity). Give them concrete examples of how the daily routines at home may be different.

You may need to prepare your family for potential problems that are frightening even to talk about. For example, one man, Jake, was discovered to have a brain tumor after he had a seizure and lost consciousness. In addition to beginning cancer therapy, he was put on antiseizure medicines. His doctors warned him of the possibility of another seizure, and instructed him and his wife about what to do should this happen. His doctor appreciated their reluctance to tell their seven- and nine-year-old sons about the possibility of their dad having another seizure.

The doctor offered simple advice: "Tell them that there is a small possibility of your having a seizure, and that this is what they should do if they are alone with you when it happens: . . . ." The couple went home and, after practicing what they would say, were very matter of fact when explaining the problem to their sons. The younger one said, "OK," and didn't seem bothered by the new problem. The older one asked questions every day for three or four days: "How will I know it's a seizure?" "What's the first thing I should do?" "What if . . . ?" "What if . . . ?" Having his questions answered calmly and methodically helped him move into his comfort zone with the new problem.

Whenever you are in doubt about the need to tell your kids something, or how to tell them, ask your doctors and nurses for advice. These professionals address these matters all the time. **It might be wise to save discussions with your children about a new problem, medical or otherwise, until you have a plan of action.** This way, your announcement of bad news can be coupled with both the realistic hope of a solution and a concrete strategy.

In my opinion, there is value to protecting your children from the full impact of some things. **Finding a healthy balance between sharing and shielding can be a difficult judgment call.** In general, if the problem doesn't affect your kids' world, try to avoid burdening them. For instance, they don't need to know the specifics of your finances or the details of your tests or medical problems. I have tried to spare my children the emotional roller coaster of false alarms and the minor complications of survivorship. A few of the times when my test results suggested the possibility of recurrent cancer, I waited until the final results were in. When the test turned out to be a false alarm, the case was closed, and I hadn't worried them needlessly.

You can't avoid the stress associated with false alarms and minor problems, but, many times, your children can. A common situation is that you develop a side effect of treatment that is not obvious to them and can be treated on an out-patient basis, such as bloody stools or impotence. In most cases, your children do not need to be informed about this additional problem and its treatment.

The overriding concern is trust and communication. Many times, you can't hide the evidence. A urine bag strapped to your leg is a dead giveaway that something has happened. Uncharacteristic short-temperedness, or phone calls that are suspiciously frequent and serious, suggest a problem in a more subtle but still tangible way. In instances such as these, explain what the changes are, what they mean, how long they will continue (as far as you know), and how they will impact the children's routines.

If you are unsure whether or not you can keep a problem to yourself without leaking clues, in most cases it is best to err on the side of talking about it. Confronting difficulties up front ensures that your kids won't draw erroneous conclusions. In many cases, after a parent has had cancer, children's antennae are up for any health-related happenings. They may be alert to signs of a medical problem that would be ignored by other kids their age. If your kids overhear bits and pieces of conversation, or see an unexplained bandage, they may deduce that you have a new, serious medical problem. Incomplete or erroneous information can cause your kids lingering, inescapable anxiety. Telling them calmly about your upcoming surgery to repair a scar or remove a port may cause them some distress, but it will most likely be short-lived and manageable.

When your children feel confident that they know what is going on in their family, they can more easily focus on the world of being a kid. **Problems don't have to be solved in order for your children to cease being distracted by them.** Children have an incredible capacity to forget about things in a healthy way if their three basic needs are met.

I must tell you an incredible true story that reminds me that my children are experiencing my illness very differently than I. In 1995 my middle daughter, Jessica, was evaluated by an ophthalmologist because her distance vision wasn't good and she was complaining of headaches. A thorough examination revealed near-sightedness, and a prescription for glasses was given. On the way to the eye center to get her new glasses, she asked me,"Mom, in your life have you ever had anything serious happen? You know, like getting glasses?" I'd been in remission only two months, after four and a half years of various cancer treatments, and she was asking me if I'd ever had a serious medical problem! I kept quiet for a minute or two, stifling the giggle inside me that celebrated her apparent disregard of the illness that so oppressed me. When she remained silent, I said offhandedly, "Well, the cancer was kind of a drag." To which she responded quickly, "Oh yeah! I forgot about that," and then chuckled. For the moment, she did forget in a wonderful, healthy way.

Providing a Framework for Dealing with the Changes

Along with the facts, whenever possible offer positive ways to interpret them. One man explained to his kids, "The urine bag is taking the pressure off my bladder and protecting my kidneys." One time, when I knew that I couldn't hide my anxiety, I told mine, "I'm short-tempered today because one of my tests didn't look right. I might have a new problem. We've got to give the doctor time to figure out why the test results weren't normal so we can know how to get me better. I'll tell you as soon as we know more."

If you can make light of an unpleasant change, you can teach your children that the mood doesn't always have to be serious even when the problem is. My husband, Ted, had been losing his hair for two or three years before my cancer diagnosis. When chemotherapy made me bald as a billiard ball, Ted and I encouraged our children to laugh about it. Ted would tease me in front of the kids, "You're balder than I am!" I would smile, wink at my children, and say, "Yes, but my hair will grow back when I'm done with the chemo." The children quickly caught on, and joined in the family joke. Neither Ted nor I minded being teased, so this approach worked for us. For others, the hair loss for the mother or the father may be too emotionally charged, and the attempt at humor may backfire, worsening the situation. Try to make sure that specific jokes will work before you use them in front of the kids. Check with your spouse, "How would you feel if we joked about your hair loss?"

When I was getting chemotherapy, my children had to know each time I had an intravenous catheter placed in my forearm for treatment. I couldn't risk them bumping it, and possibly dislodging it, when they hugged me. I also couldn't afford for them to discover it by accident, such as if they walked in on me when I was changing my clothes. Unexpected discovery would undermine our bond of trust—that I will always keep them informed.

The day of my first dose of chemotherapy, I came home and showed them the catheter. They didn't like it at all—the bandaged catheter was scary-looking and made me appear "sick." Through their squirming, I tried to help them perceive the catheter as something positive: "I'm glad I have this catheter. When the nurse puts in the catheter, I only get stuck once and then I don't have to get stuck anymore for the rest of the week. Once it's in, it doesn't hurt me at all. Without it, the nurse has to stick my arm two or three times every day."

I gently encouraged them to look at it on their own terms at their own pace: "The catheter tube is like a straw. Can you see where the straw starts? Can you follow it to where it goes into my vein?" After looking at it, they still didn't like it, but they felt the relief that comes with facing a fear. Pretty soon their revulsion wore off. They decided the catheter was "cool," and called it "the straw that keeps Mom from getting lots of needle sticks." When their friends would appear curious or disgusted, my children would take pride in being able to look at it and explain what it was and why it was a good thing.

Children Experience Stress Differently Than Adults

With all this discussion of how to manage novel or uncomfortable situations, you may worry, "Will my children be unnecessarily or permanently traumatized by seeing my incisions or baldness, hearing about the possibility of recurrence, or learning that I probably won't be able to have another baby?" Probably not, if facts are presented in an atmosphere of love and support. In general, your children's world is one of ever-unfolding wonders, rules, and feelings. If you look at it from your children's point of view, many of the household changes that accompany cancer are no different from those that accompany normal family life. Healthy moms and dads miss baseball games and ballet recitals. They also get tired, angry, and impatient, and deny many of their children's requests. To a kid, the idea that you can get a big red rash from poison ivy is just as dramatic as the notion of medicine causing baldness. Learning that you could get killed in a car accident on your way home from the store is just as scary as learning about your risk of recurrence. The sadness your children would feel about Dad missing the school play is just as painful if his absence is because he's having his appendix out, is away on a business trip, or is getting chemotherapy.

Try to distinguish your children's emotions that are due to concern about the cancer from those that are a response to the general disruption in their lives. In every home, with or without a sick parent, there are times of disappointment for children. For example, a much-anticipated family outing may be canceled. In a family where everyone is healthy, if the reason for the canceled trip is a car breakdown, the children can express sadness or anger without the parents worrying about any great implications of their mood. If your family trip is canceled because you woke up feeling too lousy to travel, your children's display of disappointment may be totally unrelated to any concerns about your illness or the bigger questions of life, even if they say something like "I hate cancer." They're just piqued that the plans fell through.

You may need to point out to them that all kids have to deal with rearrangements at home, plans that fall through, and delayed gratification. When you communicate with your words and actions that many of the disappointments they are experiencing are like those due to events unrelated to illness, the adjustment will be easier for everyone than if you always appear worried, anxious, or guilty. Your home situation definitely is not normal, but many of the problems and necessary adjustments that you are facing happen in other families dealing with the ups and downs of everyday life. Try to keep things in perspective.

This is easier to do when you remember that children experience stresses differently than adults. For one thing, children are not the decision-makers. Like all parents, you subject your children to daily surprises and demands that they don't like: you leave them with baby-sitters, potty train them, send them to school, and interrupt their games and phone calls so that they can run errands with you. These day-to-day demands or expectations cause your children stress, but rarely do they cause emotional harm. The degree of strain your kids feel is affected by your words and actions. When your children perceive that the situation doesn't bother you much, they feel it is acceptable, even if they don't like it. Your children are rebelling or expressing distress when they say, "Mom, I'm in the middle of a game. Please don't make me go to bed." The calmer and more definitive you are about the demand, the easier your children will adapt to the demand or rule.

As mentioned in Chapter 1, your choice of words and tone of voice send a message about the significance of each of the stresses in your children's lives. For example, five-year-olds who cling to their parents on the first day of preschool are demonstrating normal separation anxiety. Parents who apologize to their children ("Oh, I'm so sorry I have to leave you") or plead ("Please don't cry, I'll buy you a treat if you stay") or linger ("I won't leave until you're not upset anymore") send their children a message: "My leaving is stressful and it's not OK.

Preschool is not a safe place without me. You can't handle it if I leave you."

Parents help their children through the anxiety of this milestone separation by responding with measured reassurance: "I know you're sad about leaving me. I'll be back when both hands of the clock are on the twelve. Your teacher is excited about teaching you new games today. I can't wait to hear you tell me all about your first day when I pick you up. I love you. [hug] Bye." I suffered from my own separation anxiety when I dropped each of my little ones off at school, but drew on my sense of responsibility to keep my tears in check until they were safely out of sight. Invariably, the teacher would tell me that they stopped crying as soon as I left.

You may be having a hard time dealing with cancer-related changes even when you accept the rational argument that they are similar to other life changes. Where you may have felt fine leaving your children with a baby-sitter to go out to dinner before your diagnosis, even if they were unhappy about it, you may find yourself feeling guilty for leaving them with the same baby-sitter to get chemotherapy. What's the difference? Your children's expression of unhappiness taps into your feelings about the unfairness, stress, and loss your illness has forced on you. You elected to go out to dinner; you didn't choose to get cancer.

One mother with leukemia felt terrible about the unavoidable changes in child care arrangements during the time of her bone marrow transplant, a feeling exacerbated by telephone calls during which her children cried about how much they hated the substitute caretaker. Nurses overheard her saying into the phone, "I'm so sorry you don't like her. I'm trying to get home as fast as I can." The mother was weakened from radiation and chemotherapy and, understandably, was responding emotionally to her children's distress. Her children picked up on her sense of guilt and tried to make her feel bad enough about their predicament that she would come back home. Of course, the mother could not come home, so the telephone calls became a source of tension, anger, frustration, disappointment, and sadness for all of them.

The mother was in a compromised state and had inadvertently encouraged her children to persist in trying to get her to come home. The nurses helped her to see that the precious time and energy of the phone calls could be better used reinforcing her love or helping the children adjust to the temporary, if prolonged, separation. Since a parent's natural response is to feel guilty and sympathize, moms and dads in this situation may do well to rehearse ahead of time what they'll say.

If your treatment requires a prolonged separation from home, either hospitalization or treatment in a center far from home, you may want to try an approach such as, "I know you miss me. I miss you too. My doctors and I are doing everything we can to get me better as fast as possible, but it's going to take a few more weeks before I can come home. I'd love to hear you tell me a story about [school, baseball, or whatever]." Help them feel connected: "I look at the card you sent all the time, and it makes me feel better. My nurses can't believe you're only eight years old and can draw like that." Or, "I've asked Uncle Jim to take a video of your school play so I can see it when I'm better." *After* you hang up, deal with your feelings of guilt, anxiety, sadness, and helplessness while taking comfort in having helped your young loved ones.

Your children's anger and fear can get in their way when they are with you or talking to you on the telephone. If you find that they are too upset to be comforted by your words, it may be helpful to give them something tangible, such as a special handkerchief or a cheery note. Trinkets and written messages allow your children to feel your love later, when they are calmer. Try letting them wear one of your T-shirts or a pair of your socks to bed if they have a hard time falling asleep when you are apart.

My son had a hard time falling asleep whenever I was away from home. He needed me to take him through his nightly bedtime routine, which included snuggling, listening to him give thanks for five blessings from God that he enjoyed that day, and singing his special song. This difficulty resolved after I spent ten minutes preparing an audiotape which talked him through his usual bedtime rituals. Now, whenever I have to be away from home, my taped voice calms and comforts him as I tell him it is time to thank God for five gifts. My husband tells me how he can hear William naming his choices aloud during the minute of silence on the tape which follows my providing a few suggestions. Usually my son is drifting off by the time the tape is playing the last words of his lullaby.

The problem encountered by the woman hospitalized with leukemia is but one example of the way feelings of guilt can paralyze or misdirect your words and actions, making a bad situation worse. You did not want or choose to have cancer. **You are not responsible for the unavoidable changes and problems that accompany treatment. The best you can do is to minimize the most detrimental effects, and try to turn the negative aspects of your illness into positive learning or bonding experiences.**

Your children have looked to you to explain how rain comes from clouds, to protect them from hot stoves, and to comfort them when a favorite toy breaks. Now they will look to you for explanations, protection, and comfort about the ways illness has disturbed the order in their lives. Answer their questions about the facts and emotions surrounding cancer as methodically and lovingly as you answer their other daily questions. Children are resilient. With proper support, they are able to process, integrate, and grow through all the changes of their lives, cancer-related and otherwise.

Your children's verbal and body language can provide clues to their current need for information or emotional support. When you are talking about your medical condition, are they interested, comprehending, and asking appropriate questions? If so, keep on talking. If they are squirming, asking unrelated questions, or avoiding eye contact, back off. If your children seem to do best with less information and discussion, you don't have to tell them anything beyond the basic facts: "Dad has cancer. He's getting radiation. We expect him to get better."

There are times when information *must* be shared despite your children's resistance. "Darling, I know that you don't like talking about this, but there are some things you need to know about Mom in order to be prepared." If your children appear unable to handle information that you feel must be shared, such as a change in child care arrangements or a parent's worsening condition, get professional help.

Understanding Children's Questions

When your children ask questions or bring up topics, pay attention to why they may be seeking such information. Is it simple inquisitiveness? Or does the question reflect an emotional or physical need that is not being met? Are they confused about some aspect of your condition or treatment? Are they worried? Providing them with the most helpful answers hinges on your understanding why they are asking what they're asking. Many parents hesitate to answer at all, unsure of what, exactly, their children intend with their questions. Mothers and fathers often fear hurting their children by saying too much. In general, children (like adults) are protected from information overload by only hearing or processing what they can handle. **Too little information can be at least as damaging as too much. Don't tell them everything, but tell them enough.**

Expect to review the same information over and over and over. They will hear what you say differently depending upon their age and maturity, their emotional readiness, the specifics of the circumstances, and their experiences. As with adults, their ability to believe or take comfort in a fact or concept may lag months or years behind their ability to comprehend it. From the first week of my diagnosis on, my children all knew the correct answer to the question "Is cancer catching?" But it took a long time for each of them to believe it enough to stop worrying.

There are different levels to understanding even the simple facts. An episode with my son demonstrated in a most dramatic way that children can appear to understand or know something when in fact they don't. William was twenty-two months old when I was originally diagnosed in 1990. "Cancer" became a primary word in his vocabulary. Over the next few years, partly due to the copies of my books scattered around the house, he learned how to spell "cancer" before he could spell "cat." Yet when I flew to California for treatment of recurrent disease in 1994, William, then five years old, asked, "Why is Mom in California?" His sister answered, "To get medicine for her cancer." William then looked at my husband and asked, "Dad, what *is* cancer?" William hadn't yet grasped the abstraction that cancer was tumors inside me that threatened my health; all he knew was that cancer was something I had that was associated with me going to doctors and writing books, and not working as a doctor.

Throughout your survivorship journey, circumstances change, as do the needs and readiness of your children. **Educating your children about your illness is a process that demands vigilant attention to their ever-evolving needs. In return, you have endless opportunities to communicate, clarify, and reinforce your messages of knowledge and hope.**

Protecting Children's Need for Normalcy

To talk about what is normal touches on deep philosophical questions. "Normal" is a relative term. For me, the key to normalcy is providing an environment that helps my children grow in a healthy way. My children definitely need extra attention when dealing with the special circumstances of my illness. But they also benefit from a rhythm to their days that includes time and attention, both theirs and mine, devoted to the usual matters of childhood.

Not surprisingly, the amount of attention that should be focused on your illness, and the most helpful responses to your children's needs and wants, change over time. It seems to me that when your home life is in a state of relative equilibrium with some sense of routine, your children benefit from your efforts to steer away from health-related discussions. Children need to be children, concerned about their schoolwork and friends. A routine of chemotherapy and serial scans may be most unsettling for you, making you feel trapped in a cage of emotional stress. But your children can and should be encouraged to escape. The day my daughter asked me if I'd ever had a serious medical problem like near-sightedness, I was blown away. How could she forget, even for a moment, that I'd had cancer? I could never forget, much as I'd like to. The point here is that they *can* forget for moments or hours at a time. Forgetting is healthy, giving them the emotional space they need to grow. I'm not helping my kids by reminding them daily of my diminished appetite, or uncomfortable hip, or anxiety about my forthcoming checkup. This is not to suggest that I am a martyr, acting as if everything is fine, denying the reality of the stresses associated with my illness. Rather, **I am trying to protect the pockets of normalcy for my growing children**.

As an example, imagine that your children are going on a three-day camping trip. You may not be able to skip your cancer treatments, but when you are with them you can try to keep the focus where it would be if you weren't sick. Even though your overriding focus might be your radiation treatments next week, you can try to act involved with the preparations for your children's imminent departure for their camping trip. If it's too hard to hide your anxiety or sadness, you can tell your children, "I'm very excited about your trip and I'm really glad you're going. I may not seem very excited because [my leg is hurting, I'm very tired from the treatments, I'm nervous about my scans]."

There have been occasions when I've told my kids, "Thinking about your [sleepover, soccer game, or whatever] helps me forget about my treatments. I feel happy when I think about you having a good time with your friends." During those times when you are too compromised, physically or emotionally, to be or appear involved in your children's activities, arrange for other adults to be surrogate parents. Children always benefit when parents preserve the framework of normal routine.

When Cancer Seems to Be the Only Focus of Attention

I have found that, when I'm not careful, cancer can monopolize the spotlight, especially in my children's minds. The effect can be damaging—forcing my children to see all of life through the prism of cancer and creating a sense of powerlessness against bad times. Children dealing with a parent's cancer, especially during the treatment phase, can lose touch with how to laugh and be silly, or shift attention from something serious to something light.

There were many times, such as when I picked my children

up from school or watched them play sports, when my children wanted and needed me to be focused on them, just as I would be if I didn't have cancer. I remember too many times after my diagnosis when I was at the playground with my children and a friend or acquaintance would see us playing together and greet me with expressions of concern or a request for an update on my health. This may have been good or bad for me, depending upon my need for emotional support at the time, but their focus on my health instead of more casual topics of conversation sent a message to my children that I was, first and foremost, a sick person with fragile health. These well-intentioned displays of concern by others reinforced in my children's minds that we were living with a serious problem. Also, at those times when we had been immersed in a fun or relaxing activity, the voiced concern about my health jolted all of us back to a seriousness that was difficult to escape. What I had intended as a rejuvenating enterprise became a draining one.

I handled this by asking the parents of my children's friends to avoid asking me in front of my children how I was feeling or what was happening medically. This was not the time to worry about offending people. My true friends wanted me to get well, too, and welcomed direction on how best to help my family, even if it meant *not* doing something. My goal was to protect my children's "recharge" time. Unexpectedly, this tactic has helped keep me charged, too.

This is not to deny the seriousness of the situation or the benefits of receiving support from other adults, but to emphasize the **value of protecting the times and places where you can replenish your children's emotional stores**. Even in the most dire of circumstances, it is healthy to escape, laugh, act goofy, and just plain have fun. This is especially true when the illness is chronic. Whether there is a crisis or just the stress of ongoing treatment or recovery, you can't eliminate all the extra tension your children feel related to your illness. **You *can* help your children replenish their emotional stores and thus deal with the added stress.**

Facilitate situations where your children can learn and play, freed of reminders of your illness. The child who enjoys school and friends, and who is able to pursue his or her individual interests, will return home with more emotional energy for dealing with a parent who is ill. Reassure your children that their ill parent is well cared for while they are gone. Find the time to talk with them about school, after-school activities, and their outings with friends. When your time feels stretched to the limit by the illness, remember that an investment of ten or fifteen minutes alone with each child will pay great dividends in your children's emotional well-being now and later. **You can't afford not to spend the time with them.**

Am I depriving my children of a valuable lesson about helping others and accepting help? Not at all. My children have ample opportunity to appreciate the concern and goodwill of others who offer support, send cards and letters, deliver meals, and help out with car pooling. By guiding my friends and family to talk about topics other than my health in social settings, I allow my children to see that people can care deeply *and* treat me normally.

Am I being dishonest by acting interested in their activities during one of my hard days when I could use some attention from my friends? Your children's need for normalcy is only one of many factors that help to determine the best approach for any given day. The way to balance your immediate needs against your children's will be discussed more fully in Chapter 7.

Another common practical obstacle to meeting your children's needs over the long haul are phone calls. My telephone was both friend and foe. When I was needy, calls were a primary source of information, comfort, validation, distraction, and hope. Calling my friends and family was an effective way for me to diminish my sense of loneliness, anxiety, and fear.

Unfortunately, the ringing phone took a toll on my whole family. Keeping people up-to-date, and responding graciously to others' expressions of concern, was draining. What was news to each caller was the umpteenth narrative for me or Ted. Most calls came in the evenings, interrupting dinner, bath time, and the rituals that helped my children prepare for sleep.

Torn between wanting to stay connected to my friends and family and needing to save my time and energy for my children, I began sending updates. Once or twice a year I would compose a letter that began "Dear Family and Friends." After outlining the news about my medical condition and any plans for treatment, I would share how we were adjusting to it all. Most times, my children and I turned it into a fun family activity, doing the job assembly-line style. It was a big job to stuff all those envelopes and put on the mailing labels and stamps, but an investment well worth it. My friends and family told me repeatedly how grateful they were for the news; it was all the feedback they needed to know that their letters, cards, gifts, and prayers were helpful. My letters calmed their concerns and nourished their hopes for us. The newsletters were an important way for me to show my appreciation for their vital support.

The letters became an important way to help keep my energy account balanced. The number of incoming calls was reduced drastically. These letters were one more tangible way to show my children that we weren't alone. This afternoon, as I looked over a few of my old newsletters in preparation for writing this section, I realized that my letters have become a diary of my journey, another way for me to remember how far I've come.

There still were nights when people called me, or I wanted to reach out. When my children wanted me to help them prepare for bed, I usually could delay my call until after they had time with me and were settled in bed for the night. As long as my needs were met from day to day and week to week, I usually could postpone satisfying my short-term needs while tending to my children. On most of the occasions when I was too sick or too emotional to wait until they were asleep, Ted was able to tend to them while I took care of myself. If one of them was having a real problem, Ted or I would free ourselves to work out the trouble. If we couldn't, we would find someone whom the children trusted to fill in for us. I believe that **we have an obligation to our sons and daughters to try not to abandon them to their own thoughts when they need us or another adult.**

Handling the Crises

Routine and a sense of normalcy are healthy for your children, but there must be a balance: cancer-related issues also need attention, which may disrupt routines, especially when there is a crisis. Most times you can control when and how much attention is turned to medical or other serious topics. In a crisis, your energies will be, and should be, focused on making optimal decisions. I'll repeat what I said in Chapter 2, never lose sight of your number one priority: doing what you need to do to have the best chance of getting well. At these times, help your children to understand that you have to put your energy and attention into making important decisions that will affect the whole family. Reassure your children of your love, and let them know that your distraction from them is temporary. Until the crisis settles down, the children will have to have their needs fulfilled, in whole or in part, by meaningful substitutes such as family and friends.

If Dad is having an operation for cancer, or if Mom is in the hospital for a bone marrow transplant, cancer will be the main focus of attention in the family. Period. No way around it. To attempt otherwise would confuse the children with mixed signals. Children might think, "I see Dad looking upset when he talks on the phone or when he's doing stuff and thinks I'm not looking, but all he ever talks to me about is my homework and getting the chores done. He's talking like this bone marrow transplant is no big deal but he's acting too strange."

What atmosphere do you aim to create at home in a crisis? "Normal" is an answer that is "none of the above." Your children's world can't feel normal when there is a health crisis with a parent. You will have to talk about the illness every day to update the children and keep the lines of communication open. Strive for a new routine geared to the crisis, in an environment of love and concern for each member of the home. Keep children in the loop with the family while helping them maintain their personal friendships and interests. While overall you should try to preserve routines and enforce limits, be willing to make exceptions when circumstances dictate. This approach requires sound judgement at a time when both the ill and well parent are compromised physically and emotionally. However, knowing the challenge in parenting during a crisis makes dealing with the fallout a little easier.

Two concrete ways to help your children during a crisis are to let them participate to whatever degree they feel most comfortable, and to keep their world as normal as possible under the circumstances. Between discussions with your children about serious issues, talk about the things you would if nobody were sick. Try to keep their school and home responsibilities as close as possible to what they would be otherwise.

Routines and constancy of love and limits are reassuring to youngsters and teenagers, even when they rebel (often especially when they rebel). The more they see you functioning as their parent, the safer they feel. If, from a wheelchair or hospital bed, you discipline them verbally or help them with their homework, they will feel the reassuring sameness of your chastisement or assistance more than the differentness of your immobility. The challenge here is to **find the balance between making the exceptions that will help your children deal with the extra stresses at home, and enforcing rules and routines that comfort them with stability and predictability.** Allowing your children to stay up with Mom past the usual bedtime since Dad is in the hospital may be just what they need; other times, the routine of the normal bedtime and the good night's sleep are the best medicine for them.

Family Meetings

Family meetings can be a routine that provides another way to keep communication open, screen for problems, and work toward solutions. My family had regular meetings before my diagnosis. During these bimonthly fifteen- to twenty-minute gatherings, we'd applaud achievements, review upcoming events and changes, and discuss family problems. It was during one of these meetings years ago that my children found a solution to the bickering that always erupted before car trips over who got to sit in the coveted front passenger seat: they started a rotation of "car kid of the week." Having made the rules themselves, they accept without argument whoever is assigned for the week.

After my diagnosis, these low-key assemblies became one more place to inform the kids about my medical situation, answer their questions, prepare them for changes, and be alerted to problems. In addition, meetings encouraged cooperation, boosted family morale, and helped decrease anyone's loneliness. Discussions about the stresses were balanced by brief brainstorming sessions on ways to escape the seriousness. The children learned ways to talk about difficult issues, and shared effective coping tools with each other that complemented those that Ted and I covered one-on-one.

Calling all the children together only when you are at the end of your rope usually signals problems and tempers that are out of control. This is not an environment conducive to teamwork. One benefit of regularly scheduled meetings is that you can tackle difficulties when they are still small and manageable. Regular meetings provide some distance from the problem, and thus more neutral ground to deal with tensions between family members. Also, when cancer-related issues are discussed in the context of family meetings, your illness is normalized as but one of many family concerns, and not the overriding focus it often tends to be.

With your time and energy already stretched to the limit, you may feel overwhelmed with the thought of adding something else to your schedule. Or, if you have only two or three family members, you may feel that you really don't need a formal setting. I've come to see family meetings as one of the most time-efficient tools for delivering important information and fostering cooperation, mutual respect, and understanding. If possible, set aside fifteen minutes every one to two weeks. Meeting times should be scheduled when everyone is well fed (i.e., not right before dinner) and relatively available (i.e., not in the rush of preparing for school or during the time-slot for one of the kids' baseball practice). Take the phone off the hook or let the answering machine intercept your calls. An agenda (e.g., call to order, announcement of the past weeks' accomplishments that merit applause or thanks, old business, new business, adjournment) helps the leader keep things organized and moving. Taking notes helps separate these meetings from casual conversations at the dinner table and provides a reference for reviewing progress on old problems.

Occasionally Ted and I have to remind everyone that the purpose of the meeting is to work out problems together. We have tried to steer the children away from assigning blame or venting bad feelings without any attempt to solve the cause of the ill-will. Some authorities advise against voting, but my family has found it useful in certain situations. We usually have drawn time limits around trial solutions: "Let's try it this way for one week. At the next meeting, we'll decide if it's working or if we need to try another plan." We try to end every meeting on a positive note, be it sharing the week's funniest jokes or giving out the kids' allowances.

Doing What Feels Right for Your Family

You know your children best. If you feel that they are doing well, continue to talk about your illness openly even when well-meaning family and friends express their concern, with questions such as "Do you really need to be telling them so much about cancer? Don't you think you talk too much about cancer?" As each of your children is unique, each requires a different type and amount of information. In most cases, you are the best judge of how much information is too little or too much for each one.

You can listen to unsolicited judgments or suggestions about your child-rearing, if you want, but trust your instincts. Unlike your friends and relatives, your family is living with the reality of your illness day in and day out. You know best what is affecting each of your children and how they react. You may want to take the time to explain to your friends and relatives what you are doing, or ask one person to be the communicator. If others' advice or comments cause you to have doubts, you might want to get the opinion of your doctor or nurse, the children's pediatrician, a counselor, or other professional. Exercise your right and responsibility to do what is best for you and your children.

SUMMARY

- Recognize choices that are "none of the above."
- Adjust your priorities to fit the current circumstances.
- Imagine yourself as having an energy bank account.
- Tell each child enough to deal with his or her world and to satisfy his or her need for information.
- Try to find a healthy balance between sharing problems with your children and shielding them.
- Ask doctors, nurses, social workers, and survivors for advice when you are unsure of what to tell your children or how to tell them.
- In general, children are protected from information overload by only hearing or processing what they can handle. Too little information can be as harmful as too much.
- Expect to review the same information over and over.
- Provide a framework for dealing with changes.
- Protect the pockets of normalcy.
- Protect the times when your children can replenish their emotional stores.
- In a crisis, try to find a healthy balance between maintaining routines and making necessary exceptions.
- Encourage your children to participate to whatever degree they feel comfortable.
- Try holding regular family meetings.
- Say and do what feels right for your family.
- Continue to keep all teachers and coaches up-to-date regarding your condition and how you'd like things handled with your children.

4

Grief, Fear, and Children's Other Emotions

Dealing with Loss and Grief

Grief is the healthy, healing human response to loss. There is no pill or magic way to bypass the painful grief process. Just as you must grieve your many losses after your diagnosis in order to move forward, your children must grieve theirs. One of your responsibilities as a parent is to help your children recognize their losses, grieve appropriately, and move on.

Try not to confuse your losses with your children's. The parent with cancer sustains big and little losses—health, a sense of well-being, self-sufficiency, or a body part, to name just a few. The well spouse experiences losses too, such as a sense of well-being if he or she is overtaxed physically and emotionally, the comfort of routines, the security of the partner's presence.

As your children learn of the anticipated changes, they will become aware of their own losses, such as your companionship, and practical things such as a convenient ride to school or help with a project. Your children lose their innocence of youth to some degree or another; they don't enjoy the same sense of security that you will always be there for them or that they are invulnerable to disaster.

When you are with your children, help them face *their* losses, not yours. Children grieve a bit differently than adults, at times exhibiting any combination of disbelief, physical symptoms, regressive behavior, self-blame, fear(s), anxiety, unusual aggressiveness, withdrawal, and unusual quietness. Your children may not realize that they are grieving, let alone what they are grieving. Acknowledge your children's losses; this will help them recognize and accept them, and put them in perspective. Validate their pain, and offer comfort. Teach them how experiencing and expressing normal grief enables people to adjust and move forward after loss. **Reassure them that, although their losses may be prolonged or permanent, grief is a temporary, healing process.**

When a family trip gets canceled in order to proceed with your cancer treatment, acknowledge the loss and justify the decision: "I know how much you were looking forward to the trip but we just found out that I have to get my treatment the same week. The medicine may not work as well if I wait until we get back. We don't have any choice." Tell your children that it is normal to be sad over their losses. A canceled trip to the zoo may be an insignificant loss to you (or even a relief) but a tremendous disappointment to your kids. Encourage them to express their anger, disappointment, frustration, or sadness, feelings that are all part of the grieving process.

While guiding and supporting them through their grief, help them look for a silver lining—something good in the bad: "Dad can't play catch anymore, but he has more time to play computer games with you." Make it clear that the deprivation is still unfortunate, and deserves to be grieved, but that good things sometimes happen because of a bad situation. People seem to have natural tendencies toward being optimistic or pessimistic, but optimism can be learned and practiced.

Appreciating positive outcomes, also called secondary gains (something positive gained because of something negative), is adaptive and healing in a difficult situation. Cards, gifts, flowers, and special attention from friends and family nourish you as the shock of your diagnosis wears off and when the going gets rough. Your children benefit from special attentions, too. No matter how well everyone is dealing with the situation, your children are dealing with added stress at home. Going to the movies with friends can give them a boost when Dad is in the hospital. Staying up extra late on a weekend night to watch a television show with Mom and Dad rewards them for handling the stresses and losses sustained during the week. Give them good feelings and memories to balance the bad. This must be done within a framework of maintaining the normal routine, rules, and discipline as much as possible. How much is too much special treatment is a judgment call based on what your children need at the time.

You may worry that your children are enjoying the extra treats too much and are taking advantage of the secondary gains. A child who feigns a stomach ache in order to stay home from school on the day of an exam is taking advantage of the secondary gains in the negative sense of the word. So, too, is the child, home with a legitimate illness, who demands lots of toys and bosses everyone around. This is different from the sick child who enjoys the distraction of daytime cartoons and the attention of lunch on a tray that is decorated with a little flower. In this latter case, the little pleasures ease the discomforts of the temporary illness.

Encouraging your children to look for some positive outcomes of a negative situation teaches them that **they have some control over their perception of the world**. Looking for and appreciating positive outcomes makes life easier, happier, and more meaningful. "Something good from something bad" has been an important theme in my home. Just as my children remind me to buckle up in the car, they remind me to try to look at the good side of a bad situation if I appear especially sad or disappointed. **This is not an attempt to deny unpleasant feelings, but a way to begin to transcend them.**

What about the times when you can't find any apparent good in a situation? Clearly, there is nothing good about Mom losing her fertility when she wanted more children or Dad losing his voice box. In these difficult situations, make it clear that the change or loss was worth it, and as painful as the sacrifice has been, you would make the same treatment choice again.

Dealing with Fear

Fear is the feeling of anxiety and dread you experience in the face of a perceived threat. Fear will try to be your ever-present companion throughout your cancer treatment and recovery. As a parent, your fear of the future—your future and your children's—can immobilize you. Only after learning how to tame your own fears can you help your children deal with theirs.

Their fear of your death, the first fear that may come to your mind, is discussed in the next chapter. Many other anxieties can threaten your children's healthy adjustment: fear that you pose a danger to them because of either your disease or its treatment; fear of being abandoned by you or rejected by their friends; fear that your condition is worsening; and so on.

Find out what fears your children harbor. Their fears based on misinformation are easily dissolved with simple clarification of the facts. However, until your children understand and believe the explanations, their fears may persist. For example, your children may be afraid that your radiation treatments make you radioactive and therefore dangerous to them. First, validate their fear as a legitimate and understandable one before you try to reassure them with accurate information. Then address the forces that generate and perpetuate the fear. In this example, your necessary isolation during the administering of radiation therapy may lead them to believe that you are radioactive. Explain that the technicians can't come into the room while you are being radiated because *the machine* will radiate the technicians, too. Emphasize that your children don't have to be protected from you. Give examples that support your point: "When I stay outside in the sun too long, I get sunburned. You can touch my skin and be close to me, but you won't get sunburned from me. You can only get sunburned from the sun." Since it's hard to prove your point, repeated reinforcement over time will help them believe the facts and tame the fear.

Some of your children's irrational fears can be allayed simply through the act of giving them voice. Concern that you won't love them if you are bald or that you will forget them while you are in the hospital are unfounded but real. Help your children subdue these fears by encouraging them to talk about them aloud. Listen, then validate their feelings as normal and understandable. Offer reassurance that you will not let what they fear come to pass. At the very least, your **being aware of their specific fears will help you understand their behavior**.

When fears persist, teach your children techniques for taming them. Make up stories, songs, or rhymes with a happy ending to the situation or problem they fear. Self-talk is another useful technique for taming normal fears. Teach your children to talk positively to themselves to help them deal with a worry: "I can't get cancer from Dad, I can't get cancer from Dad." Even when they don't yet believe it, self-talk is comforting. *The Little Engine That Could* is a classic tale of positive thinking. Challenged by the steep grade of the mountain, the little train chugged, "I think I can, I think I can, I think I can!" So, too, your children can talk to themselves, helping to build faith in things that are true but hard to accept.

Inspire your children's hopes and courage with visual reminders. Draw a picture of The Little Engine That Could, and hang it in their room. Ziggy is a key player in the cheering squad of my home: he hangs on the wall in our kitchen, saying, "Keep on Hoping, Keep on Wishing, Keep on Believing!" When my kids say, "I can't . . . ," I point to the poster. They pass his picture of encouragement every time they walk through the kitchen. Even I'm not too old to be encouraged by Ziggy. During the seemingly endless months of postcancer fatigue, his words have helped me believe that I can get well. If your kids reject your suggestions for projects that nourish optimism, do them for yourself. At the very least, your efforts may help you feel better, which will help them. At best, **your children will see the value in shaping the environment to foster hope and courage**.

Instill confidence that they can deal with new and frightening situations: "I know that you're scared to go to the hospital, but I believe that you can get used to all the machines and tubes after a few more visits." Praise your children when they take a step toward taming a fear; build their confidence in facing and dealing with frightening and anxiety-provoking situations: "You showed real bravery today. Yesterday you weren't ready to go to the hospital, and today you were able to go and wave to Mom from the door to her room. That's great. Mom told everyone how happy she was to see you." Provide role models through stories of real or fictional children who surmount hurdles similar to the ones yours are facing. Hearing the story of a child who overcomes his dread and is glad to be united with his ill parent may help to inspire your child who has a similar anxiety. Take your children's fears seriously. **Never laugh or tease about their fears or talk about them to other people in front of your children.**

Although looking problems in the eye is important, **distraction from the crisis or chronic health issues is adaptive, too.** Temporary diversion can help adults and children deal with a frightening thought, or with a situation that isn't going away soon. Find challenging, upbeat activities or entertaining company to take your children's minds off their worries for awhile. They can't be worried about you at the same time they are laughing at a funny movie.

Even if you do everything right—listen patiently to your children's fears, validate them, offer a zillion explanations and reassurances, and teach different tricks for taming them—your children's fears may persist. As exasperating as it may be for you, already stretched by the demands of your illness, try to hang in there and address their fears for as long as they persist and whenever they recur. Even when you think to yourself, "I thought we took care of that a long time ago," the same fears can resurface as your circumstances evolve and your children mature.

Sometimes children are too afraid to deal with the fear. That was William's problem when he was five: That summer a Texas storm headed our way. I started to review the "tornado drill" with our children, reminding them where they were supposed to go and what they were supposed to do if they heard the siren signaling a tornado sighting. My son clapped his little hands over his ears and interrupted me, crying, "Stop! Stop talking about tornados. Talking about them makes me scared." I took him in my arms and said, "William, of course tornados make you feel scared. They scare me, too. But if you cover your ears and don't learn what to do, then every time we have a storm you're going to feel as afraid as you feel right now. If you learn what to do when you hear the siren, from now on when a storm comes you won't be as scared, because you'll know what to do to stay as safe as possible."

Cancer and many of the attendant changes are frightening. The world *is* scary in many ways. Children must learn healthy ways to live with the risks and real problems that they face every day. One of your jobs is to teach your children to tame their fears. If your children don't want to hear your suggestions for taming their fears, maybe it will help to share the story of William who was afraid of tornados, a little boy who became less afraid after learning what to do.

Helping Children Work through Negative Emotions

Grief and fear are only two of the many unpleasant emotions children may experience when a parent falls ill. Anger, sadness, loneliness, resentment, frustration, depression, and disappointment are some of the many other emotions that occur. A vital ingredient to mental health is learning about feelings and how to express them in adaptive ways. While you are under treatment, your children may experience a multitude of feelings in succession, or even at the same time. Your children may seem highly emotional. There is a legitimate reason for them to show intense emotions—you have a life-threatening disease. As excruciating as it may be for you to witness their distress, you must encourage your children to express their negative feelings about what is going on in their lives. **Expression can be therapeutic by itself, and through the help it summons.**

Depending upon your kids' ages and past experiences, they may not even recognize their feelings for what they are. **Teach your children that unpleasant feelings such as anger and disappointment are normal and expected.** Reassure your children that their emotions will lessen with time, and cannot affect your medical outcome. Be clear: "It's OK to be angry. Your feelings will not make me more sick."

Your children's emotions are not the problem but rather the reaction to, or signal of, a problem. Open communication is indispensable to working through feelings, especially those that are uncomfortable, confusing, or embarrassing. Offer your children frequent and varied opportunities to talk. Nonverbal outlets are important, too, especially for children who need to loosen up before talking, for quiet children, for kids who are too young to express themselves well verbally, and for children whose emotional relief is incomplete even after a good discussion. Give suggestions for safe and healthy ways to relieve bad feelings. This approach will help them through this crisis and lay the groundwork for dealing with future problems that cause negative feelings.

When your children are upset, frightened, or disappointed, they will have these feelings whether or not they find a safe place to express them. Without such release, their uncomfortable emotions persist, coloring their perception of the world and affecting their personal growth. When your children share their painful thoughts and feelings with you or another adult, they can begin to transcend them. At the very least, they will feel understood and can begin to let go of negative emotions that drain their resources. More likely, they will learn through the cancer-related stresses and losses how to survive adversity.

You may find it distressing to hear their pain, especially if they direct the negative emotions at you, or their emotions are intense. Steel yourself. Only if your own emotions don't get in the way can you help your children work through difficult times. Prepare yourself to absorb their outbursts or to let ugly comments slide off your back. What can you do when your children's angry words—"You never do anything with me!"—scorn your heroic efforts? Trying to defend yourself with responses like "That's not true. I do spend time with you, even when it's really hard for me" often won't help when your children are feeling angry. Indeed, at that moment they may feel that you are never available. Their emotions can distort their perception of reality.

Validate your children's feelings: "I can see that you are furious" or "It's understandable that you feel angry." That may be all that is needed, but more often this is just a start. If your children's sense of abandonment persists, addressing the problem further may help them out of their anger. At a calmer moment, point out the times that you have done things together. Work together with your children to find new ways to help them feel that you participate actively in their lives.

The facts may reflect that you are giving your children every ounce of energy you have and that you spend good quality time with them every day. If, in this setting, your children cry that you are never there for them, one of two things is probably happening: Your children have discovered that cries of abandonment push your buttons, make you feel guilty, and effectively draw you away from any other activity. Or they honestly feel abandoned despite your attentions. Instead of denying their claim, learn why they feel the way they do. Take an honest look at the times you spend with your children. Are you focused on them, or are you with them physically but concentrating on other things? Do you need to review again with them the reasons why you have to take a long nap every day or why you have to go to work?

Children sometimes feel more emotional about everything when a parent is ill. Larry, a six-year-old whose mother was undergoing chemotherapy, looked at his broken toy and whimpered, "I'm so unlucky." His mother's outstretched arms were rejected and he ran to his room crying, "Even my mommy is broken!"

Larry felt justifiably sad and lonely, and let his mom know in a powerful way. Larry's mom rose to the occasion and was able to put her reactions aside while she addressed her child's feelings. First, she acknowledged them. Then she tried to help Larry see the situation differently: "You're sad because your toy broke. Maybe you're sad because your broken toy reminds you that I'm sick, and you don't like that at all, so you feel even more sad. Let's see if we can fix this toy [or buy a new one, if that is reasonable]. If we can't, let's look for something else fun to do."

Throughout my journey with cancer, there have been times when my own children were hurting emotionally. Seeing them unhappy, angry, frightened, or anxious about my illness has been incredibly painful for me. I wanted to be the strong, understanding, patient mother they needed me to be. Unfortunately, the same problems that precipitated my children's distress—say, that I needed another operation or more treatments, or I was too sick to attend a school event—necessarily were circumstances that drained my reserves. During crises, I was physically and emotionally compromised, making it difficult for me to keep things in perspective or respond well to my children. The last thing I wanted was to deal with a child who was acting out.

As an example, the day after being diagnosed with recurrent cancer in 1992, as I was saying my good-byes before going to a support group meeting, Rebecca, then seven years old, screamed at me, "You're leaving? You care more about your cancer and your support group friends than me!" Emotionally and physically exhausted from the two days of tests and bad news, I needed to go to my support group.

Seeing her distress stung more than usual—sad and afraid, I overempathized with everyone else's emotional displays, especially hers. At the same time, self-preservation instincts pushed me to keep a distance from her pain. To mix up the emotional goulash further, I was angry at her for acting out when I was so vulnerable and in need myself. She was making me feel like a bad mother.

I left my angry daughter in the loving care of my husband and in-laws and calmed my reflex emotional reaction to her explosive accusation by repeating my mantra over and over: **"We're doing OK if she feels safe enough with me to express herself."** I reminded myself that she was dealing with the bad news, too. Her reaction was a mirror reflection of mine. We were both upset, disappointed, angry, sad, and afraid. Certainly I was upset about my recurrence; why shouldn't she be? When she gave voice to her feelings, she helped me to know how she was doing and how I could be there for her.

Another thought that helped me absorb her anger was the

fact that I have never been ambivalent about inflicting short-term physical pain on my children in order to get them well, such as when they need vaccinations or stitches. When my toddler son was in the emergency room for intractable diarrhea, the nurses had a hard time starting the IV in his little veins, which were collapsed from dehydration. As much as I hated to see him stuck repeatedly with a needle, I knew that he had to experience the brief pain if he were to get well. My discomfort at witnessing his pain was far overshadowed by my confidence that the benefits were worth it. **When dealing with my children's emotional health, I try to accept the evidence of their short-term emotional pain as an inevitable part of their getting through a hard time.**

There is a fine line between accepting problems related to short-term adjustments and ignoring signs of a deeper, longer-term problem. If you are unsure, discuss what is happening with your spouse, your nurses, or a counselor.

Underlying all of your words and actions must be the sense that you do not feel sorry for yourself or your children. Pity helps no one. If you do not feel sorry for yourself or them, your children will be less likely to get stuck in self-pity. We can learn something valuable from children with illnesses such as diabetes or spina bifida, or injuries such as amputations. Those who are not pitied by their parents grow up to be more self-reliant and to have greater self-respect than those whose parents feel sorry for them and spoil them. Children do not naturally feel sorry for themselves; self-pity is taught.

There are times when it takes conscious effort to refrain from a posture of pity for your children. This is especially so if you are feeling sorry for yourself. Instances of sympathy or indulgence are fine, as long as they are relatively brief and infrequent. Recognize when you are slipping into a mode of pity for your children or yourself, and train yourself to avoid it. Life doesn't come with a problem-free guarantee. Teaching your children to see your cancer as a fact of life, rather than as an unfair punishment, encourages them to grow up with a capacity to adapt to all the big and little bumps on the road of life.

After a Bad Day

We all stuff our feelings away and put on our "professional mom or dad" hat when our children need us immediately. But like moms and dads everywhere, with or without cancer, sometimes I blow it—yelling back at my child who is having a rough time "I can't listen to your whining anymore! Go to your room." When I'm really flustered or burned out, I'll scream something very intelligent like "Stop acting like a baby, it's just a paper cut!" or "Your room better be clean in five minutes or you'll never have another sleepover in your life!" Of course, once I've calmed down, I can't redo the scenario in a healthier, more productive way. So the next best option has been to apologize to my children and make sure they understand that I was having a bad time.

One of the many Harpham family rules is that we always forgive each other for having bad days. There have been monstrous times, with a huge gulf between how I wanted to react to my children and what actually happened. Out of the ugly times came wonderful lessons for my children: **we always love each other even when I can't "be there" for them; we're not perfect and that's OK; we always try our best, but the best we can do is the best we can do—sometimes the best isn't so terrific, and that's OK too.**

Another family rule is that we forgive *ourselves.* Everyone makes mistakes. Lots of them. And it feels bad, too. When you make a mistake, you can try to ignore it, in which case you're likely to make the same error again and again. Or you can beat yourself up about it, feeling sorry for yourself. There is another option, as illustrated in my favorite part of *The Lion King.* It is worth watching the whole movie for the few minutes where Rafiki, the wise mystic baboon, teaches Simba that **it is what we do after a mistake that determines whether the overall experience is debilitating or energizing**. Rafiki smacks Simba, the Lion King, over the head with his cane. Simba asks, "Why did you do that?" To which Rafiki answers, "Why worry about it? It's in the past." Simba, still rubbing his sore head, says, "Yes, but it hurts!" Rafiki in his wisdom explains, "Simba, you can spend all your time feeling bad about it, or . . ." and he swipes his cane again at Simba's head. This time, Simba ducks, avoiding another blow. Rafiki continues his sentence, ". . . or you can learn from it."

Using Play as a Therapeutic Tool

Another way to help your children work through their feelings is through a basic tool in child development: therapeutic play. Depending upon your children's ages and levels of understanding, play allows them to express and understand emotions such as fear and anger. They can test ideas, release tension, and heal emotional hurts. Play-acting with medical equipment and dolls may encourage discussion of medical issues that your children find too threatening or embarrassing to talk about when addressed directly. Dolls or puppets can be used to teach your children about operations, catheters, incisions, and other medical interventions. For young or anxious children, using a stuffed animal may feel less threatening or frightening than using a human figure. Many hospitals teach children about blood tests and IVs this way: "Let's take Brown Bear to the hospital to have his blood checked."

Offer to play a game of "chemo," "radiation," or "operation." Let your children choose which role they play. You or your children may be more comfortable playing a more generic game such as "school" or "house," where concerns about abandonment, illness, death, disability, or love can be verbalized and acted out. Role-playing offers one more window into how your children are doing. Playing with your kids can be the most important work you do.

Drawings are another tool to help you know what's going on in the minds and hearts of your children. Suggest drawings: "Draw a picture of things that make you happy [or mad or sad]." Be very general. Let your children lead. Without any training, you can use their drawings as a springboard for discussion: "Tell me about your picture. Tell me about the little boy in your picture."

Your children's art work may signal brewing trouble. Pictures with sad or angry faces may reflect your children's inner emotions. "I hate Dad" or "God is dead" scribbled in books indicates distress. If discussion with your children about the art work or scribbles is not reassuring, or if discussion is impossible, consider reviewing the drawings with a professional such as your children's pediatrician, the hospital social worker, or a counselor.

Use play, also, as an escape from the seriousness of your illness. Uninterrupted play with your children sends valuable messages: I want to be with you; I enjoy being with you; cancer is not the only important thing in my life; it's good to laugh and have fun even in the worst of times. If you played with your children before your illness, the familiar games will be comforting and reassuring. If you didn't, the new activities will be a positive outcome of a negative circumstance.

When Laughter Can Be Healing Medicine

Proverbs 17:22 teaches, "A joyful heart makes for good health, despondency dries up the bones." Humor can be therapeutic. Families that tease and make puns need to keep these tension-relievers working. Jokes about cancer can be particularly healing. Sit-coms and stand-up comedians are popular because they make fun of the areas in our lives that cause stress, confusion, disappointment, and a sense of powerlessness. Favorite topics for stand-up comics are parenting, government, pollution, violence, and sex—all issues that are serious, close to our hearts, and subject to many forces beyond our control.

Where can you poke fun at cancer? You can start at home. Laughing together at your weaknesses and fears is therapeutic because it offers your family an escape from confining circumstances, a lift away from depression or despair, and a new perspective to balance the negative forces and feelings. Humor is one way to demystify the undefinable, disempower that which is feared, and regain a sense of control. Laughter encourages healthy acceptance and forgiveness, two forces that help bond the family together. As we say in our house, "A no-hair day is better than a bad hair day!"

You may find it impossible to have a light heart in hard

times. Joking about cancer may fly in the face of your deepest feelings. Even if you are usually upbeat, you may feel too burdened or sad to rustle up some fun. Try to do it for your children. Joking around with them, for their sake, will help them and may lift your spirits, too. Other times, the only way humor will help you is with the awareness that you are teaching your children a good lesson to carry with them for times of tribulation.

Humor can backfire if your children are trying to communicate serious thoughts or feelings and you respond with a funny story or pun. This tells them that you don't take their concerns seriously or that you don't want to address the real issues. It can also backfire if it triggers bad feelings in you that adversely affect your behavior with your children. Ill-timed comic relief can lead to disaster; well-timed, it can heal.

Dealing with Regression and Failure

Even with the benefits of play and humor, your children's behavior may regress during crises. You can expect to see some behaviors that had been abandoned long ago. Your four-year-old suddenly is wetting pants, thumb-sucking, or using baby talk. Older children may be neglecting chores or homework, playing with toys or dolls from earlier years, or shadowing your every move. **These unconscious signals of increased needs are normal, and usually resolve when the children feel that their three fundamental needs are being addressed.**

You may find your children's behavior annoying or frustrating. With my children, I found it helpful to remind myself that they were not *trying* to be bad or cause more trouble for me. **They just needed a little extra loving, and some time to adjust to whatever stress prompted the regression.** Once they became comfortable with the new routines, I could expect the troublesome behaviors to resolve.

Unfortunately, sometimes I felt really angry when, suddenly, they couldn't brush their teeth or dress themselves anymore or drink a glass of milk without spilling it. That meant more work for me! And it tapped into my sense of frustration about not being able to control my illness, making me feel as if I didn't have control over *anything*. To help keep myself from overreacting, I had to consciously remind myself that **when they lost ground, my emotions about other things could too easily be misdirected against them.** And ranting and raving never helped anything.

Respond to a drop in your child's performance quickly. With younger children, try to spend some time with each one alone, free from distraction. That may be all you need to do. If the problems persist, you may have to be more firm. Make it clear that you will help them through the crisis so that they can regain the lost ground and move forward again.

You can try a straightforward approach, such as saying "Johnny, you stopped wetting your bed at night last January. For the past couple of weeks, your sheets have been wet in the morning. Do you have any idea why?" **Acknowledge the behavior in a supportive way so your child does not feel ashamed.** Work through what might be causing it with your child. In the case of bed-wetting, you may discover that your child has a urinary tract infection. Or you both may realize that your child has been drinking more fluids in the evening; heightened separation anxiety may have increased the frequency of your child's requests for a glass of water during the night. If there is no obvious cause, you can offer the possible connection of the backslide to the changes at home. Offer reassurance that the ability to stay dry at night will return.

In summary, evaluate regressive behavior with an open mind; don't punish regression; encourage progress; arrange daily private time; and be patient. If, with time, things don't get any better, ask for advice from your child's pediatrician or a counselor.

With older children, you may or may not want to acknowledge the evidence of regression. They could feel very embarrassed if you pointed out behavior that they consider babyish. As with younger children, spending time with each child alone and giving them an opportunity to talk may be all they need. In a nonthreatening way, you can mention that rules are responsibilities that must be met, such as by saying "I haven't had to put your dishes in the sink [or bring your lunch to school, or whatever] for a long time, and lately I've had to pick up after you. I know it's been hard with Dad sick and everything, but it's important that we all keep doing our jobs."

I tried always to use this composed approach, but of course there were many times when I did lose my cool. Within minutes or hours, I'd calm down, apologize to the victim of my little tantrum, and resume the patient, supportive approach.

Regression is one possible reaction to stress. Failure is another. When school grades drop or behavior is unsatisfactory, sit down with your children and address the problem directly. In a gentle and supportive way, explain, "I see that your grades are lower than usual. I know that Dad being sick is distracting you and making it harder for you to do as well in school as you usually do. Your job is to keep up your grades. In fact, one of the ways you can help Dad is by keeping up your work; then he can concentrate more on getting better. Let's figure out what we can do to help you get your work done well even while he's sick." Failure is not an acceptable response to a crisis.

Try to understand what is keeping your child from performing. A child answering "I'm just not interested in it now" could indicate fatigue from sleeping poorly, or concentration difficulties due to depression, anxiety, or unresolved anger. Fatigue and painful emotions are problems that need to be addressed. You may discover a practical difficulty that is easy to resolve. Your children's complaint "We don't have time to get homework done because we're always visiting Mom at the hospital and trying to keep the house clean" may indicate a time-management problem that is easily solved with some rearranging of the schedule. Get help with the housework, or let it go. Arrange visits with Mom that allow your children to get their schoolwork done. Let them know that you will help them do the best they can under the circumstances. **It's fine if the best is not their usual, but it has to be their best.**

Failure is a cry for help and should never be allowed to continue unaddressed. Friends and family may push you to "Be easy on them. They're under stress." That's like saying "Of course there's a fire alarm ringing, there's a fire!" and then leaving the fire to burn. You wouldn't ignore the ringing. Nor would you shut off the alarm without responding to the smoking blaze that triggered the alarm. In a similar way, the sensible response to your children's failure is to address the cause of the decline while offering emotional support. Don't ignore the alarm, or try to turn it off without attending to the problem(s) that set it off. Your response helps to determine the long-term outcome of this experience. Do you want to teach your children that it is OK to fail if they have a good excuse? **Children can learn how to do their best no matter what the obstacles.**

Knowing Whether Children Are Doing OK

In the midst of all this turmoil, how do you know if your children are OK? The biggest clue that your children are having trouble is a *persistent* change in behavior. You may notice a major shift in sleeping or eating patterns. Or they may appear uncharacteristically quiet, withdrawn, aggressive, violent, sad, or mischievous. Their developmental progress may seem to slow down or stop, as discussed in the preceding section. The appearance of new, more intense, or persistent fears is another clue.

Warning signs themselves are not the problem; they signal one. If your children stop eating, your job is not to get them to eat without understanding what is causing the food avoidance. Forcing your child to eat won't begin to solve the underlying problem and rarely helps the eating problem, either.

Trust your intuition about your children's state of mind. If you sense that something is amiss, you're probably right. Open-ended questions are less likely than specific ones to cause their own problems. Asking "Is anything wrong?" offers them the opening to talk about your illness if they need to but does not bring illness to mind if that isn't the source of the problem. Asking "Are you worried about my new chemotherapy?" is fine if your children were indeed already worried about it, but invites your children to begin worrying about your treatments if they weren't.

Be careful not to project your own worries on your children. There may be times when you feel insecure about your future, and your child's edginess may suggest that she is suffering from the same anxiety. Closer questioning may reveal that your child is irritable simply because you won't let her stay up late to watch a television show, or you accidentally shrank her favorite T-shirt.

This brings us back to the same monotonous conclusion: open communication will help you know when there is a problem and will help you work through it. If you are not sure if something is a problem or just normal "growing pains," get help. Even a licensed child psychologist might at times need an impartial observer to help sort out what is going on and what can be done. Parenting is tough work.

SUMMARY

- Help your children to recognize loss and to grieve.
- Help them look for a silver lining (appreciate positive outcomes of a bad situation).
- Help your children tame their fears.
- Help your children to express their emotions in a supportive environment.
- Teach your children that unpleasant feelings such as anger, disappointment, and fear are normal and expected.
- Separate your emotions from your children's.
- You are doing OK if your children feel safe enough with you to express themselves.
- Ask forgiveness; forgive yourself.
- Avoid pity.
- Find times for laughter.
- Respond to regression as a sign of stress and increased needs.
- Failure is a cry for help that must be addressed.
- Teach your children how to do the best they can *under the circumstances.*
- If you are unsure whether your children are OK, get advice from a professional.

5

Helping Your Children Live with Uncertainty and Tame Their Fear of Death

Dealing with Uncertainty and Faith

The cancer experience can expose the illusion of control held by so many. Your health has been threatened, and your future feels less certain. During treatments, it may be difficult to make plans because you aren't sure how well you'll feel from week to week. At the end of treatments, there is uncertainty about how long it will take for you to recover. Until a certain period of time has elapsed, it may be hard to feel confident that you are out of the woods with your cancer. For some types of cancer, there is no length of time after which your risk is as low as the general population, but even if you are pronounced "cured," you may feel anxious about developing cancer again or another illness, or suffering an accident. Your brush with serious illness made you aware of the uncertainty of life in a new way.

A crucial step in healthy survivorship is learning how to live with uncertainty about the future and an inability to know all the answers. You may be amazed at the sophistication of some of your children's questions. Try to keep your responses honest and as simple as possible. Sometimes the best answer will be "I don't know." In this case, let them know if you will try to get them an answer, or if they have asked the unanswerable. **This honest acknowledgment of the limits of your knowledge teaches children that they can live with unanswered questions and with uncertainty.** This may be a rich opportunity for sharing your thoughts and feelings about spiritual faith.

Teaching children about God and faith is complicated business. Like everything else, children's concept of God, and sense of faith, evolve as they mature. Their understanding of their spiritual selves will help shape their perception of the events around them. Conversely, their life experiences will affect their beliefs. Depending upon how you have explained yourself, and how they understand what you've tried to say, future circumstances over which you have no control may deepen or shatter their faith. For example, one family taught their two children that God is all powerful, and encouraged them to have faith in God's power to heal their dad. When the young father subsequently died, both children felt betrayed by their family and their God. Unresolved anger kept them from hearing any explanation other than that there really was no God. They felt alienated from the rest of the extended family who seemed to derive so much comfort from their faith. This was not what the family meant or wanted. Parents need to be aware of how their children hear their messages, especially in light of the family's circumstances.

Vulnerability is another element of life that is brought into focus by the cancer experience. Your trial of fighting feelings of vulnerability can be turned into an opportunity to teach your children how to stop fretting about things that might or will happen in the future. Worries about the future are real and normal. However, being able to focus on the present, and not future events that you can't control, is adaptive and healthy. Your children may find it easier to understand this idea if you relate it to their day-to-day experiences. Remind them of some past worry, such as how they worried about a rainy day spoiling their upcoming swim party. Review how you made backup plans and then tried not to worry anymore, because you'd done what you could do to ensure a great party and you couldn't control the weather. "Once we made our backup plans, it wouldn't have helped anything to keep worrying about the rain."

Let me share a wonderful story that saves me from crippling worries about my future: A mother took her young boy to the local station for his first train ride. The train was due to arrive in twenty minutes. Excited, the boy asked, "Do we have our tickets?" His mother answered, "Not yet. We'll buy them from the conductor after we board the train." The boy became anxious, and persisted, "Can't we buy the tickets now?" The mother answered calmly, "There's no ticket booth here. We can only buy the tickets on the train. All we can do now is wait." Each time the boy asked about the tickets, the mother told him calmly and definitively, "You'll get your ticket when you board the train." She encouraged him not to worry about the tickets and, instead, to watch all the interesting people while he waited. After a while, he accepted that he couldn't get the tickets until the train arrived. No longer worrying about them, he was able to enjoy watching the other commuters.

This story offers an allegory for letting go of worries about something that might or will happen in the future: "I can't buy a ticket until I get on the train." In many life situations, you can't do anything to prevent a problem from happening or figure out how to adjust to it until the difficulty actually arises. It's absurd to try to put a cast on an arm before it is broken. Survivors can't do anything to treat recurrence or a medical complication until the disease is a measurable reality.

Worrying or feeling sad about the possibility of recurrence won't help you feel less anxious or sad if it happens; it just increases the amount of time that you're stressed and unhappy. In response to the awful fear and anxiety I've felt every time I've learned of my own recurrent or progressive lymphoma, I've explored ways to prevent the unpleasant emotions that accompany bad news about this darn disease. The obvious answer is that I can't. If next week or next year I find out that I need more treatment, I'm going to have a lousy day. There is nothing I can do now to make the bad news easy to hear in the future.

The most I can do for myself is **learn to let go of the worry so that fear of the future doesn't steal an otherwise healthy today.** When I find myself worrying, I remind myself, "I can't buy a ticket until I get on the train."

The futility of anticipatory worrying is only one of the lessons illustrated by the train ticket metaphor. In addition, the story offers a positive response to the despair that feeds anxiety about the future. When I feel the fear of recurrence, and think, "What would I do if another lymph node popped up? I just couldn't deal with lymphoma again," I am calmed with the belief that if the cancer recurred, then I'd find the strength. While I'm in remission, I may feel that I don't have the fortitude to face recurrence again, but "I'll buy a ticket when I'm on the train." Those with a religious orientation may interpret this as "God will provide when the time comes." Your children will grow strong knowing that **people are resilient, and are able to rise to occasions beyond their imaginations**.

Dealing with the Possibility of Recurrent Cancer

Living with uncertainty is one of the tasks of survivorship. Parents can decrease the magnitude and breadth of their children's sense of uncertainty through thoughtful answers. Children inquire about a parent's illness directly or indirectly. With time, you will learn to recognize why your kids are probing, and therefore what they are unsure about. In 1991, tests confirmed that I was in complete remission after completing six cycles of intensive chemotherapy. After we told my children the glorious news, my daughter asked me, "Can your cancer come back?" She was not just curious about the biology of non-Hodgkin's lymphoma. Further discussion revealed that she was concerned about her needs being met if I got sick again.

The emotional "mother" side of me wanted to protect her from experiencing fear of recurrence, a powerful and painful emotion. The patient side of me was absorbed in the process of trying to allay my own fears of facing cancer again. At that vulnerable moment, I wanted to respond, "No, you do not have to worry about it coming back."

However, I needed to deal with reality, not wishful thinking. **Children remember what they are told.** Without hesitation, I sat down and wrapped my arm around her shoulders as I told her the truth, with my focus on hope and reassurance: "Everything looks like the cancer will not come back. If it does, it probably will be a long, long time from now, when you are grown up. But Jessica, if it comes back sooner, I will be treated again. We got through it once and we can get through it again. I'm just glad I'm better now." Her question answered and her concern addressed, she never showed signs of wondering or worrying about me having a recurrence. Our conversation provided me with a powerful new way to tame my own fear of recurrence: "I can be treated again. I handled it before and can get through it again."

Before my one-year checkup, much to everyone's surprise and chagrin, I was diagnosed with recurrent cancer. In the midst of the emotional upheaval of that terrible time, I was relieved that I hadn't lied to Jessica about the possibility of recurrence. She would have had difficulty trusting anything I said if, after my original treatment, I had reassured her that I was cured. Without this trust, she would have been deprived of the comfort and reassurance I was able to offer her during this new crisis.

My answer was truthful and helpful: recurrence was possible and her needs would be met no matter what happened. In general, if your children don't ask about recurrence, directly or indirectly, you probably don't have to bring it up. However, if your children ask about the possibility of recurrence, or if your type of cancer has a high probability of returning, tell them the truth, with the focus on hope and reassurance: "It probably won't [or may, or probably will] come back, but I can and will be treated again. I'm just glad that I'm fine now."

Hope without expectations. Talking about recurrence, whether in depth or briefly, as a possibility, will not diminish your chance of cure or your sense of hope for getting well and staying healthy. Your rational mind can accept the possibility of recurrence, even know that it is a high likelihood, and at the same time have genuine hope that you will be or are cured of the cancer. My acceptance of the fact that this type of lymphoma has a stubborn tendency to return coexists in my mind and heart with my genuine belief that I can get well. Even after my recurrences, I can talk about how we will treat the next recurrence while at the same time believing that I can fall on the good side of bad statistics and never have another recurrence. Children, too, can know grim facts and at the same time believe in the possibility of good outcomes. If anything, it is easier for children because they have fewer memories of others' unfavorable outcomes working on their subconscious.

When conversations with my children weave their way around the concern of recurrence, my response is four-pronged: I acknowledge the possibility, reassure them that we'll deal with it if we have to, reinforce both the fact that I am fine now and my belief that I can stay well, and urge them to live as if we know I'm going to be fine. **Although my specific response varies according to the mood of the conversation, the framework for my answer is steadfast.** With time, this tactic has become automatic. Sticky questions are no longer emotionally draining; now they are occasions to reinforce my belief structure for my children and myself.

Here's an illustration of the melding of frightening facts and genuine hope: I was shopping with my daughters, and we passed a woman who was wearing a fancy scarf on her head—double-layered with a braided border. She looked lovely, but we could tell she was bald. My daughter Rebecca commented, "Your scarves weren't that nice." I responded, "I know. They make much nicer ones now. If I'm ever bald again, I'd definitely buy some of those neat new scarves. Of course, I don't want to be bald ever, ever, ever again [we laughed] and I'm planning on keeping all my hair for the rest of my life [again we laughed]. But if it happened . . ." Rebecca interrupted me and, with her index finger pointed at me and her eyes squinted in an exaggerated expression of threat, said, "You'd better not get cancer again!" Her cue triggered my reflex response, "If it comes back, we'll deal with it. For now, I'm planning on never needing chemo again."

This way of talking about the uncertainty of my future health works well for my family. Our reality is that I've had a type of cancer that is notorious for returning again and again. My prepared responses to questions about recurrence serve several functions. First, my comments address my children's present concerns, and thus offer immediate comfort. Second, the repetition reinforces our faith that we can survive cancer again, if we have to. Third, it encourages closure of the conversation without blocking further discussion, if needed. Fourth, having ready answers minimizes the drain on me of dealing with an otherwise emotionally charged issue.

In my case, the probability and reality of recurrence forced me to find adaptive ways to handle this topic with my children. For other parents with cancer, the chance of recurrence is low after successful treatment. In these cases, the fact that recurrent cancer is possible but unlikely can be presented with great confidence.

Helping Children Face Their Fear of Death

Cancer is a life-threatening disease. Even though very few people are at death's door at the time they are diagnosed, most people have thoughts about dying when they learn they have cancer. For parents, especially those with young children, fear and anxiety about death are compounded by concerns about their offspring. It helps to be prepared with what you'll tell your children about death and cancer because questions and concerns about your possible death may arise anytime after you break the news of your diagnosis, no matter what your medical condition.

First and foremost, **cancer is an illness, not a death sentence.** Since your children's life-long associations with cancer will be influenced by their personal experience with your condition, they will benefit from frequent reminders that your case is not like all others. Your approach to talking about death and cancer will be defined by the many factors discussed in earlier sections on talking frankly to the children about cancer, including your medical condition and prognosis. Your spiritual beliefs play a major role in determining what you say.

How can you help your children deal with your illness if your prognosis is grim? Thinking about how to handle this troubling circumstance will help every parent dealing with cancer, even those with a fairly good prognosis. If your doctor believes that your condition will deteriorate soon, knowing what to say and do will help your children cope with their imminent loss.

Some of you may have treatment options, and may be doing well now, but have a type of cancer for which there are no known cures. Although you don't have to prepare your children for any final good-byes right now, playing through the worst case scenario in your mind, and figuring out how you would manage it, serves a few useful functions. It can help tame the fears and anxieties that arise when something makes you wonder, "What would happen to my children if . . . ?" And it may prompt you to take some steps now that you would want done should your condition worsen, such as updating your will or arranging for family and friends to participate in the many spheres of your children's lives in your absence. Even if you have reason to expect a permanent cure, mentally running through the "What if I die from this?" routine may help you by putting your brush with mortality in better perspective, and helping you tame your fear of recurrence.

One of the best favors I received when I was first diagnosed was a visit by a colleague who was a ten-year survivor himself. He came to my hospital room the afternoon I learned of my disease and the proposed treatment. After pulling up a chair, he sat down and tilted it onto the back two legs. Without spending much time on chit-chat, he proceeded to the heart of the matter: "Wendy, what's the worst thing that could happen?" Still stunned from the morning's shocking news, I started to cry. "That I die and my kids don't have a mommy." Gently, he took me through the projected horror, asking me what would happen at each step. Yes, they'd be sad and miss me. Yes, there would be people to change their diapers, feed them, take them to preschool, and put them to bed. Yes, there would be Ted and extended family to love them. As we moved from the concrete to the intangible, I was reassured by my own belief that Ted would do a good job raising them alone, if that's what he had to do. Our children would be loved and cared for, and could grow up into solid citizens without me.

That twenty-minute exercise prevented my ultimate fear from expanding in my mind like a goblin that lurks under a child's bed. From the outset, we'd drawn limits on the fears: my children would not be abandoned to shrivel up without food or loving. As awful as my dying would be, they would survive, and thrive in the long run.

My friend was confident that he could help tame my fear in one short visit because he knew Ted. What about patients who don't have a "Ted," such as single parents or those in unstable marriages? The solutions may not be as obvious or satisfying, but **even suboptimal but realistic plans will help tame the fear.** You may need guidance from extended family, friends, or professionals. Ideas for ensuring your presence in your children's lives are discussed later in this chapter. Helping children of unstable marriages is discussed in the next. Taming this fear as much as possible will free you.

Knowing If You Are Dying Now

If you would describe your prognosis as grim, what exactly does that mean? Unless you are literally on your deathbed with no potentially effective treatments available to you, you have good reason to hope for temporary or long-term improvement. As long as you have hope, you have time to live and be a parent.

Time is relative. If you learn that people with your type of cancer usually live a year, you may feel that you are dying right this minute. You are not. A year can be a long time. To children, especially young ones, twelve months seems like an eternity. Your children grow and mature in measurable ways every month, let alone every year. Take comfort in the knowledge that a short time to you is a long time relative to your children's growth. During the weeks or months of your illness, your children can learn to deal with the many changes and losses associated with your worsening condition, an education that will help them accept and deal with the possibility or reality of your death. A year from now, your children will be more mature, more prepared, and better equipped to overcome the loss than they are today.

Understanding Children's Conception of Death

Whenever mentioning or discussing death with your children, it is vital that you appreciate how children's conception of death matures developmentally, and where each of your children is in the process.

Recognize that there are no set guidelines to a child's psychological development when applying the following generalizations to your own children. It is well agreed that infants and children under three are incapable of grasping any abstract concepts such as death. However, it would be wrong to conclude that they are oblivious to the goings-on during periods of anxiety about death or mourning, and therefore don't need special attention at these times.

Children three to five, in general, perceive death as temporary and reversible. They often construct degrees of death, temporary states of death, or see death as a lesser state of being alive. Using euphemisms for death such as "went away" or "went to sleep" only compounds the difficulty. These references may arouse anxiety about normal trips or separations, or going to bed at night. In addition, very young children see themselves as the center of the universe and feel responsible for everything that happens around them. Belief in magical thinking keeps this perception intact. Children's television shows also contribute to children's perceptions of death by presenting characters who are killed repeatedly in episode after episode.

Children six to ten usually see death as an eventuality, but the result of outside forces, such as accidents or an act of God. Consequently, they often look for motive for the perceived punishment such as the parent or child having done something wrong. Death may be personified, for example as a dead man or an angel of death, as a means of defending against the realization that life may end. Bogey-men or other figures may be envisioned to keep the angel of death at a distance. This age group tends to see things in all-or-nothing terms. Usually by age ten to twelve death is perceived as an intrinsic, universal part of life, a process which culminates in a perceptible disintegration of bodily life.

Breaking the News of a Poor Prognosis

What should you tell your children when you first learn that your chances of long-term survival are slim? They need to know only that you have a serious illness, that you want to get better and are doing everything possible to get better, and that you will keep them informed of how you are doing. If you have *any* chance for improvement, you have solid ground for hope, a force that is life-enhancing if nourished by you and your children. If the hoped-for but unexpected recovery doesn't occur, your children's experiences with you over the weeks or months of your physical decline will prepare them for the possiblity or reality of your death.

If your children ask, "Are you dying?" answer them realistically but hopefully, "I'm not dying right now. I'm getting treatment that we all hope will make me better. If it doesn't work, hopefully there will be a new treatment to get me better. We will tell you if I'm getting worse, but I'm not dying now." If they ask, "Are you going to die from this cancer?" you can answer truthfully, "I don't know; I hope not. This is a bad kind of cancer to get. Many people with this kind of cancer die, but some people get better. I'm doing everything possible to be one of the people who does great." Your children may feel upset (why shouldn't they?) until they adjust, but they will have a better chance of handling whatever outcome transpires if you talk to them truthfully. Acceptance of a likely demise can coexist in our minds with hope for the unlikely recovery. Show your children how to **hope with acceptance.**

Understandably, talking about your own death is extremely difficult. You may be having trouble adjusting to the news, and may need time to figure out how hopeful you are or want to be. After being diagnosed with cancer, many people realize that hope is not all or nothing, and that they can take steps to find and nourish genuine hope. Hope is a complex force that is affected by innumerable factors: the statistics for your type of cancer (it is easier to be hopeful when the chance of cure is very high); past experiences with illness (it is easier to feel hopeful when you've seen others with cancer survive unfavorable odds);

how you feel, physically and emotionally (it is more difficult to feel hopeful when you feel sick or have pain); and how much hope your family, friends, and health-care team project (it is easier to have high hope when, clearly, your doctors do). Like any complex phenomenon, hope is dynamic, fluctuating over time. Until you have a chance to realize how much hope you have, and how much hope you want to sustain, err on the side of sounding hopeful.

The seemingly straightforward question "Are you dying?" prompts spiritual and philosophical questions. Facing your mortality challenges your beliefs about God and fate. It feels different to ponder your own death when you are a healthy person than when you have a life-threatening disease with little hope of recovery. Beliefs that calmed your anxieties and answered your questions satisfactorily when death was an abstract concept may or may not provide the same relief now. Whatever you believed before, your illness most likely will make you see or feel things differently, and it may take time before you know what you believe now. Your faith may be renewed or strengthened, shaken or lost. You may find faith in a higher being for the first time. Or you may confirm your conviction that there is nothing beyond the earthly existence that we can appreciate with our senses.

Your understanding of the mind-body connection also plays into your sense of hopefulness and the way you answer your children's questions about your prognosis. You may feel tremendous pressure to "have a positive attitude," and worry that acknowledging the possibility of death will seal your sad fate. Friends and family may try to encourage you with their belief that your recovery is in your power, if only you will believe. This confidence that the self-healing potential of our minds can control our health is supported by many popular self-help books.

I see this pervasive myth as a double-edged sword. It is good when it encourages people not to give up prematurely, and to take steps that increase comfort, personal growth, and the chance for recovery. Belief that the mind controls the body is dangerous when it makes people feel guilty for having gotten sick, and guilty again when they don't get well. As a physician and survivor, I have come to believe that the mind plays an important role in our health and quality of life. But **the mind only *affects* the body, it does not *control* the body**.

If having thoughts of death caused a fatal outcome, nobody would survive cancer. Having so-called negative thoughts—fears of illness or death—means that you're human, not despairing, unless you focus on them, or interpret them to mean that you are hopeless. Let me emphasize again that you can accept the possibility or likelihood of death in the near future and at the same time nourish hope for an unlikely recovery.

In order to help your children, you must figure out your own comfort zone regarding your level of hopefulness, spiritual beliefs, and understanding of the mind-body connection. If you can **respond to their questions and concerns realistically yet hopefully**, do it. If you can't, and it may be a while before you can, then defer to someone who shares your beliefs, such as a close family friend, relative, or member of the clergy.

Your children may want to know details about what will happen to you if you die, or to them if they die. Answer with an eye toward understanding the need behind the question. Are they truly curious? Are they fishing for further information about your current condition? Are they insecure about their own needs being met? Validate their question as acceptable and ask them why they want to know, or if there is anything special they are wondering or worrying about.

If you answer them with silence or a change of topic, you are sending a message that it is unacceptable to talk about death. Their questions can be a wonderful prelude to your teaching them to see death as a natural part of life. **If they sense that they can talk with you about death, they will have the security and support of a loving ally in one of their most difficult life tasks—facing the reality of death.** This is a blessed time to share your thoughts and feelings about God and faith. If you can't talk about death now because of your own emotional state, deflect the question to someone who can handle it for you.

What do you say to your children when, indeed, death is very near with little likelihood of improvement? There is going to be enormous pain for the children whatever you say. For a parent to have a terminal illness is a catastrophe for any child. But a greater tragedy occurs when children are shielded from the truth about their parent's condition. As heart-wrenching as the process is, give your children the chance to say good-bye, and help them find healthy ways to anticipate and cope with their loss. **You can't control the medical outcome; you can affect the impact of this life-altering loss on those loved ones you leave behind.**

Misleading your children, directly or by omission, denies them open communication, the guidance they need to find ways to anticipate and deal with their loss, and the comforting closure of saying good-bye. Since trust is vital to our most important relationships, in addition to the difficulties accompanying the premature loss of a parent, children who have been misled during a parent's terminal illness can grow up with handicaps stemming from difficulty trusting the world around them.

One twenty-nine-year-old mother downplayed the seriousness of her cancer, banking on the 95 percent likelihood of complete recovery and cure offered at the time of her diagnosis. Her three daughters (nine, seven, and four years old) were encouraged not to worry about what was happening during her treatment because "Everything will go back to normal after treatment is done." Her reassurances proved false when her brief remission ended with evidence of aggressive recurrent disease. Her family unraveled. A series of medical complications and heroic procedures preceded her premature death. The children were kept away for the last two months of her life in hopes that she would recover and they wouldn't ever have to see her looking so sick. After her death, these children had to adjust to a loss that they did not understand and for which they were totally unprepared. They will grow up having never said good-bye to their mother.

You can't protect your children from change, loss, and pain; you can help them face and deal with these facts of life in healthy, healing ways. The mother above could have shared the reality of her prognosis at the time of her original diagnosis without causing them undue anxiety by saying something like

"Most people with this type of cancer get all better and never get it again, but a few people don't get better. I want to live a long time. I'll tell you if things aren't going well, but my doctors and I are planning on me doing great."

The equation changed at the time of her recurrence. Now her chance of cure was significantly smaller, and she was going to be sick for a long time before she got better, if she were to get better. There are ways that she could have encouraged her children's understanding of the situation, while at the same time nourishing their hope. For instance, she could have said something like "Some people with recurrence of this type of cancer get all better and other people don't. We're doing everything possible to get me better and I'm planning on getting better. If the treatments aren't working, I'll let you know."

The equation changed again when, clearly, the cancer was progressing despite ongoing treatment. The fact of the matter was that she was dying. **Hiding this truth did not protect her children from the pain of their imminent loss. Rather, the healing power of normal grief was delayed and hampered because they were prevented from putting some closure on their relationship with their mother.**

This story is especially sad because this mother was loving and responsible, and tried to do what was best. The father, too, didn't want to harm the children in any way. They both believed that talking about death, especially during her original round of chemotherapy, was an unnecessary stress for their children. According to the statistics, 95 percent of people in her situation would have gotten away with the omission. Most people who go without fire insurance get away with it, too, but that doesn't mean it's a good decision to forgo buying the extra coverage.

As this story demonstrates, whenever you discuss death, the emphasis will change according to the specifics of your circumstances. **When cure is probable, the possibility of recurrence and death can be simply acknowledged and set aside. When death is likely, children do better if the probable outcome is stated clearly.**

Matching Your Degree of Optimism with the Facts

As explained in earlier sections, telling the truth and nourishing hope are not mutually exclusive even when a parent has little chance for recovery. If anything, being truthful frees your children to have hope because they are not trying to sort out the conflicting messages of what you say and what they see. When death is near, try to prepare your children for the likely ending while encouraging them to hope for an unexpected reprieve, be it days, weeks, months, or even years. Hope is life-enhancing.

When dealing with your children, it is important to match your degree of optimism about your prognosis with the facts about your disease. **Your children must be clear on the difference between what you hope will happen and what is likely to transpire.** Your children could suffer a sense of confusion, disappointment, and abandonment if they misunderstand your expressions of hope as statements of fact. For example, if you said, "I hope to get completely well," they might hear, "I know that I'm going to be completely well."

Sometimes a problem arises because of your coping style, or your belief structure. If you can't be honest with your kids, let someone else tell them what's going on. Let me share a story that will help illustrate when this would be appropriate. One young father had a serious, life-threatening cancer with little hope for improvement. He understood his prognosis, but found that he coped best by talking as if he knew that he was going to be the one out of a thousand who was cured. The adults in his life automatically used many different sources, such as their medical knowledge and past experiences, to interpret what he said and find their own level of optimism about his situation.

His children, on the other hand, were depending on him to show them how optimistic to be. His illness was their first experience with cancer. This thoughtful father put his children's interests over his own when talking with them. He modified his coping tool of unrelenting optimism against the odds, and told them that his chance of surviving was slim but that he had genuine hope that he could survive. He pointed out the distinction between could and would. Acknowledging the seriousness of his situation is something he wouldn't have done if it had not been for his children. At the same time he encouraged them to have hope.

After that discussion, he had to wrestle with his own stirred-up fears and anxieties. It took a lot of time and energy for him to get back to his comfort zone of denying the possibility of a bad outcome. He and his wife realized that his needs and his children's were incompatible. For the next several months, he continued his stance of unwavering optimism. His wife would keep the children's view reality-based with comments such as, "You know how optimistic your dad is about everything, and how hard he wants to fight this. His last scans don't look good, the tumors are still growing. Dad knows it, too."

The opposite coping style—unfounded pessimism—can also cause trouble for children, even when it works well for the parent. There are people who approach every challenge in life predicting failure. You know the type: the brilliant kid who goes into finals saying, "I'm going to flunk it," and then aces it; the track athlete who insists she'll do badly, and then comes in first in every event. As adults, these people approach everything, from fixing the sink to interviewing for jobs, with a public announcement of intent to fail. This is a variation on magical thinking, which decreases their anxiety when they feel challenged.

A straight "A" student who pulls this pretest behavior is annoying, at worst. The stakes are higher when a parent with cancer predicts problems or bad outcomes: most children will feel helpless, and have difficulty being hopeful. **Unfounded pessimism only fuels children's fears and anxieties; it does not help them deal with their present, nor does it prepare them for their future.**

If your disease is probably curable, make this clear to your children. If you are struggling with negative thoughts, share them with someone who can help you find and nourish an optimistic attitude to fit your optimistic prognosis. **When dealing with your children, always share the optimistic prognosis given you by your doctor.**

You may have disease that, right now, is not life-threatening, but also is not likely to be cured. Until new, more effective treatments are discovered, you are looking at disease control (keeping persistent cancer at bay) or repeated remissions and recurrences. Tell your children so, with the focus on hope. You are not helping your children by being overly optimistic about being cured. You may find it useful to shift that optimism to hope for an unexpected cure with current treatment or hope for the development of a cure in your lifetime, as well as living as fully as possible within the constraints of your disease.

Obviously you need to know the facts in order to tell your children the truth about anything in a realistically hopeful way. If you are still unsure of the facts, tell your children that you don't know, but you'll find out. Sometimes they'll ask the unanswerable. As noted before, saying calmly, "I don't know," teaches your children that it's OK if you don't know everything all the time, that you can find out what you need to know, and that meanwhile you can live with the uncertainty.

Preparing Children for Your Death

It's almost impossible to face your own mortality without stirring up powerful emotions. Try to separate your own emotions from the emotions your children are experiencing. Don't confuse your sadness over dying prematurely with your children's sadness over losing their parent. Your job as a parent is to address their feelings and concerns, which are very different from yours.

Your own death represents a final good-bye to everything you hold dear. Death means leaving your family, work, play, body, and the earthly world. The most painful loss may be your inability to complete the most important life task you have—raising your children. To your children, your death means something very different. They are losing the comfort and security of your presence.

When your remaining time is likely to be very short, your children can be encouraged to accept the terrible news and take advantage of what little time is left. Guide your children to help you feel comfortable and loved, and to share meaningful good-byes. If your children resist, don't push them to say or do things they are not ready or willing to do. Try to find some way that they can connect with you within their own comfort zone. Acceptance, so difficult for adults, is perplexing for children. Seek the assistance of a professional counselor or member of the clergy to help your children make the most of these last times with you. You'll never be able to recapture this time; goodbyes are memories that your children will carry with them through the rest of their lives.

One of the problems associated with losing a parent prematurely is the risk of your children feeling abandoned. You can minimize this risk by making clear the distinction between *leaving them*—separating physically because you have no choice—and *abandoning them*—choosing to forsake them. Make it clear that you are doing everything reasonable to survive, even when that means being sick from the cancer or its treatment. When there are no effective therapies left, help your children see your impending death as your release from pain and untreatable disease. Help them know that you would never abandon them.

You can help your children prepare for and deal with their loss by creating a memento to leave behind. I knew a devoted young mother who, dying of metastatic ovarian cancer, was devastated by the realization that she would not participate in her son's life as he grew up. Unwilling to be denied her role, she dictated letters to him: one for each birthday until he was twenty, and others for the day he graduated high school, for his wedding day, for the day he became a father, and so on. In her letters, she shared her hopes and dreams for him at each of these milestones, and the love and pride she felt for him. Her husband typed the letters for her, sealed them in envelopes, and gave them to their son on the appropriate occasions. This woman's presence survived in his life long after she had died. The son enjoyed the comfort and joy of her love, feelings that far outshone the yearning for her triggered by the letters. He would have missed her at these times anyway.

You may feel unable to write letters because you find it too disturbing to have your feelings about dying brought so close to the surface. One way to overcome your anxiety and grief is by comparing writing letters to buying fire insurance: you are uncomfortable when you buy the policy because you are looking at the possibility of disaster. Once you've bought the policy, though, the uneasiness subsides and now you can enjoy a sense of satisfaction because you are covered.

I know of a widower who suffered a major heart attack when he was in his forties and his two sons were under eight. Sure that he had less than six months to live, he started writing a journal to his boys. In it, he shared his deepest thoughts and feelings about them, his illness, and his life. He was tremendously comforted by the knowledge that he had made it possible for his children to know him better and understand what had happened in the event of his premature death.

Amazingly, he survived and recovered completely. He has been sharing his thoughts and feelings in person these past twenty years since his heart attack. His journal became a rich gift, a vibrant reminder of the intensity of his love for his children when he was threatened with his own death. You cannot protect your children from the loss and pain of a parent's terminal illness; you can help your children face and deal with their suffering in a healing way.

If you don't enjoy or feel comfortable writing, you may find it easier to dictate your thoughts into a tape recorder or make a video. Or you may ask a close friend with whom you've shared your thoughts and feelings to write a letter to each child for you. Don't feel guilty if none of this works for you. These are just a few of many options that can help you tame the grief, fears, and anxieties you feel about leaving your children.

If religion is an integral part of your family, allow your religious leaders to help prepare your children for whatever lies ahead. Clergy can not only help guide your family throughout your illness, they can help nourish your children's spiritual footing into the future. Be sure that the leaders you enlist share your family's philosophy of life. Even within the same denomination, beliefs can vary widely.

I will conclude this chapter with the inspiring story of one woman who was living proof to me of the power of love. The afternoon that I met her to lend her a copy of the latest version of this manuscript, it was obvious that her remaining time could be measured in weeks. Diagnosed two years earlier with cancer, this mother of three had undergone a bone marrow transplant and had achieved remission. Four months later, the disease recurred with a vengeance and was not responding to any treatment.

With tears in her eyes but determination in her voice despite choppy sentences broken by gasps for breath, she told me about the grief counseling she'd arranged for her children. A search through the yellow pages and a call to the hospital social worker had led her to a wonderful local counseling service that specialized in children's grief. She'd already taken her family to the center so that the counselors could develop some rapport with the children as well as learn more about her. She was doing whatever she could to make sure that her children would receive expert, loving care after she died.

This woman was the embodiment of what I feared most, especially because I was undergoing chemotherapy at the time of our meeting: a mother who wasn't going to live long enough to finish the job of raising her young children. Yet since our visit, I have been less afraid. That afternoon, she gave me more than the invaluable tip of checking out grief counseling services during the terminal phase of the parent's illness. She showed me that **the same motivation that helps us endure difficult treatments can help us prepare our children for living on without us.** She gave me confidence that whenever my time comes, I can do what I have to do.

SUMMARY

- Learn how to prevent fear of the future from stealing an otherwise healthy today.
- Talking about recurrence will not diminish your chance of cure.
- Nourish hope with acceptance.
- Talk through the worst case scenario.
- Time is relative; you can do a lot of parenting in a few weeks or months.

- Learn where each of your children is in their understanding of death.
- Sharing the truth of your terminal state allows you to help your children find healthy ways to anticipate and cope with their loss.
- Try to match your degree of optimism with the facts.
- If you can't talk about death, let someone who can do it.
- Prepare something to leave for your children.
- Find out about grief counseling for your children.

6

Family Members with Special Needs: Teenagers, Single Parents, the Well Spouse

Caring for Teenagers

Teenagers are notorious for the challenge they present to parents. Sometimes they seem so grown up, and other times so immature. They can demonstrate apparent personality changes, or seem to reject all that their parents hold dear. An understanding of the world of teenagers will help you know how to help yours. What distinguishes this stage of life is that teenagers separate themselves from their family and test out their independence. In regard to applying the principles and philosophies of dealing with cancer, your teenaged child sits on the fence—sometimes having the exact same needs as their younger counterparts, other times fitting more into the adult category, and occasionally juggling problems and feelings unique to this age group.

Understanding Issues Common to Teens and Youngsters

During the normal process of breaking free of the family unit, your teenagers benefit from your feedback or reassurance.

Home is usually a safe place to test out new ideas or values. Even your disapproval can be comforting because it provides stable boundaries and a sense of constancy in a time of exploration and dramatic change. Difficulties can arise during your treatments because of your decreased availability to your kids. Without your feedback, they may be more timid about venturing out, or may have difficulty weathering the setbacks, disappointments, and times of confusion characteristic of their age.

Teenage behavior can regress under the more ordinary stressful circumstances, such as when they suffer a falling out with close friends, or fail to make the baseball team. The nurturing offered in the privacy and safety of your home helps them heal their wounds and resume their ventures while saving face in their social circles. Your teenagers, stressed by your illness, can exhibit regressive behavior at home and at school. This backslide serves the same function it does for younger children: it signals their distress and need for extra attention. However, teenagers may be more self-conscious about reverting to behaviors that they want to leave behind. They may be made to feel self-conscious by adults who reprimand them for acting like a baby. Regression is normal, adaptive, and usually temporary if your teenagers' needs are subsequently fulfilled. Either don't make a big deal of the regressive behavior, or go ahead and reassure your adolescents that it's OK to act childish sometimes. The important point is to **respond to their cry for extra attention and nurturing, so that they can regain lost ground and move forward.**

Teens, like little kids, feel disappointed, sad, or angry when promises are broken, such as when you can't attend an important event or a vacation gets postponed. The priority of the ill parent's treatments is obvious to you, but can get blurred in the more self-centered world view of your teen. Intellectually they can understand, and be expected to grasp, that their sweet-sixteen party is less important than their parent's operation. In most cases, they realize that the loss is not earth-shattering in the grand scheme of things.

The tension arises when they *feel* that their world has fallen apart, and when they feel that the party cancellation is a loss that is enormous, unfair, and sometimes unnecessary. You can help your teen through these unpleasant emotions by acknowledging them: "I know that this is a tremendous letdown for you, and you're mad about it. We had no choice about it; if we did, we would have tried to make it work." Help them find safe ways to vent their feelings. Steer them toward realistic choices for alternative solutions, and when possible, focus on the advantages or bonuses of these choices.

Children of all ages can find it unsettling to have rules and priorities shift at home. Try to continue to enforce routines and rules with your teens, making exceptions when dictated by the circumstances. Part of the work of the teenage years is challenging rules and limits. The changes and uncertainties accompanying your illness make it all the more important to maintain reliable routines and limits at home. Routine is reassuring. Also, a close rein protects teens from acting on emotion-driven impulses that can get them into trouble.

Magical thinking is expected in little children, and is hard to undo. Don't assume that your teenagers are too old for magical thinking. The tooth fairy and Easter bunny may have been exposed long ago, but rituals such as wishing on stars often remain a vital part of their day-to-day life. In particular, **your teens may harbor feelings of guilt or anxiety that somehow their thoughts caused your illness**. They can feel responsible in some way for your medical outcome. The likelihood that they have memories of using inflammatory language, or doing something offensive, is high given teenagers' natural drive to challenge or fight their parents.

Belief in magical thinking can be extremely damaging in this situation. One approach that usually works is to be direct: "Lots of people worry that they caused something bad to happen to someone else even though that would be impossible. For example, a gymnast might hope that her rival gets sick and can't compete. If, by chance, the competitor does get sick with strep throat, the thoughts did not cause the illness; the strep infection caused it. So, even if you've been angry with me and had evil thoughts, there is no way that anything you did, or said, or thought made me sick."

Magical thinking can exert a powerful tug against your teenager's powers of reason; repeated reassurances help to curb the negative impact of magical thinking. Directing your teen's ritualistic behavior toward more adaptive thoughts and activities may help. For those who believe in a higher being, prayer can be a comforting and spiritually enriching way to focus thoughts. Encouraging them to do something positive can help counteract anxieties that they wish(ed) you ill. Concrete ways for them to reinforce their belief that they want you well include activities such as making or buying greeting cards, or helping with a chore. Point out that these activities can help make you feel better, even if they don't help you get well.

Loss is a major issue for every family member when a parent has cancer. Teenagers, like their younger siblings, experience and express grief differently than adults. Allow them to grieve in their own ways, and help guide them toward healthy grieving styles when they seem stuck.

Lastly, teenagers need to be told the facts over and over. They need for their worries and concerns to be heard again and again. Their need for comfort and reassurance is ongoing. Their emotions and maturity will evolve over time, as will their readiness to hear things and their ability to process information. As with the younger ones, repeat, repeat, repeat. Err on the side of going over things too often; they will let you know when they don't need it.

Understanding Issues Common to Teens and Adults

Crises can cause a restructuring of responsibilities at home. Adult tasks may fall on the next best person—your teenager. Teens vary in how much stress and responsibility they can handle. Yours may rise to the occasion, accomplishing things that they may not have known they could do, and feeling proud and fulfilled in the process. Or yours may feel pushed into situations for which they don't feel equipped, and therefore be destined to fail. Feelings of inadequacy worsen, and self-esteem plummets. They can't make things better and, in their failed attempt to help, they may have made things worse. The obvious solution is to be sensitive to which tasks are being assigned to the teenagers, and to how well they are holding up. Talk with your teens: "You've had to take over some adult jobs. How's it going for you? Is there anything I need to know, any way I can make your jobs easier?" From a practical point of view, circumstances often lend themselves to letting adult responsibilities fall on the teenager's shoulders, such as making meals, doing housework, carpooling, or helping the younger children with their homework. It is crucial to remember that **teenagers may have the bodies of adults but they have the hearts and minds of teenage *children*.** They can mow lawns and drive car pools, but it is unwise and unfair to everyone to ask or expect them to take the ill parent's place in nurturing or disciplining younger siblings, or listening to your concerns. **A teenager cannot be a substitute mom or dad to the younger siblings or surrogate spouse to the well parent.** Get help from other adults whenever possible, even if you have to ask for it.

When cancer treatments are prolonged, or a parent is chronically ill, the risk of teen burnout is great. A few extra responsibilities that are tolerable or even exciting in the short run can be draining over a long period of time. Teens can get run down by the additional demands, but feel guilty about complaining. Feeling trapped, they may perform their usual or added tasks poorly. Or everything may continue to be completed as requested, but they appear depressed, angry, frustrated, or overwhelmed. The advice offered later in this chapter for preventing burnout in the well spouse also applies to the teenager who assumes extra responsibilities.

Try to be alert to what your teenagers are giving up, such as time with friends or time to do homework. Whenever possible, protect them from sacrificing too many things that offer them sustenance through hard times. Try to avoid having the needs of the ill parent always take precedence over your teen's need to "just hang around with friends." Carefree time with buddies can rejuvenate your teenagers, offering either escape from the seriousness of home or an opportunity to share the burden with friends. You may feel that you can't afford for the teen to be off with friends when there is so much to do at home, but **you *must* allow respite time for your teens.**

As teenagers flex their muscles of independence and control over the world, they may want to take on the parent's illness as their own fight. When the parent and teenager share similar outlooks and styles, this can be affirming. If your teenagers seem irritated by the parent undergoing treatment, see if they are unhappy with the way the parent is handling the illness. You can start with an open-ended question such as "You seem annoyed with me at times. It would help me to know if there is something I'm doing (or not doing) that's bothering you." If no revelations are forthcoming, you can try again: "Different people all have their own ways of facing a challenge. How do you feel about the way I'm dealing with my illness?" If you hit pay dirt, and find that your kids are disappointed or unhappy about your approach, let them share their thoughts and feelings. Then explain to them that this is *your* illness, and you get to choose how you face it. Help them to let go, and accept your way of coping. If you don't hit pay dirt, try again and again.

Like adults, teens benefit from sharing their feelings in a safe environment where they feel understood. Your teenagers may find it difficult to talk with their friends about your illness. It's worth calling the hospital or the local office of the American Cancer Society to find out if there is a support group for teenage children of patients. If your teenagers resist going, you may suggest that it would help you, and that they can stop going if they don't like it after two visits. If there are no age-appropriate support groups available, you may consider arranging a visit with a social worker. Again, assure your teens that they can stop going after one or two visits if it's not helpful.

Helping Teens Stay Connected to Their Friends

Another aspect of cancer that is similar for teens and adults is the feeling that "strange and strained" has become the norm at home, a feeling that is heightened by daily interactions with the well world. Very young children spend most of their time at home, and therefore are insulated from the contrast between their lives and those of people who are not dealing with illness.

Teenagers may suffer more than adults from this feeling of being different because conformity and connectedness with peers take on great importance at this stage of life. The pain of feeling different is heightened when other kids are cruel. Class clowns and bullies, or schoolmates who are insensitive or awkward, may make fun of your teen's obligation to baby-sit a younger sibling. Find out if your teenager is struggling with this. Help your teen do what feels right, and handle the fallout.

The demands at home may make it difficult or impossible to reciprocate with friends, thus alienating teens even more from their friends. If your teenagers can't invite friends over, or are too embarrassed to do so, they may be at a disadvantage in the peer pecking order. In addition, many of their friends and acquaintances may be inexperienced at dealing with illness and so may feel unable to relate to your teens' issues, and may ignore or avoid any mention of your illness. Some of your teens' friends may try to minimize contact with you, not so much because they fear catching it (which may be the case with some) but because your illness threatens their sense of being indestructible. Teenagers feel bulletproof, and don't want that sense of immortality threatened.

You can help your teenagers by listening to how things are going for them with their friends, and helping them understand why things may be awkward. Your teenager may be more accepting of a friend's apparent rejection if he or she understands that the friend may be afraid to talk about your illness, fearful of saying the wrong thing. You can guide your teens to help their friends feel more comfortable mentioning your illness. You can show your children how to tell their friends that it's OK if they avoid talking about the illness, but not if they avoid being together. Remind your children that the situation is temporary. Maybe share with them when you've faced similar reactions from your friends, and how you have handled it. Try to encourage your teenagers to stay connected with their contemporaries. Sometimes the best you can do is to preserve their times with friends when serious personal matters are unlikely to come up, such as practice for a team sport.

Understanding Issues Unique to Teens: Drugs, Alcohol, Sexuality, and Dangerous Behaviors

Parents of grown children often reflect that raising their children through the teenage years was their most challenging and frightening time. Drugs, alcohol, sex, curfews, and choice of friends can be volatile topics even in homes where the parents feel comfortable discussing them and their teens overall feel happy and secure. In some families, the parents are having their own problems in these areas, or they avoid bringing them up simply because they don't feel comfortable with them.

The lure of illicit drugs or alcohol, promiscuous sexual interactions, and dangerous behaviors such as speed-skating without a helmet, or running across train tracks as close to an oncoming train as possible (the game called "chicken") can be stronger for teens who don't feel secure at home, have poor self-esteem, or are undergoing serious stress such as that related to a move to a new city, divorce, or death in the family.

The risk of your teenagers pursuing these activities is real during your illness because, in addition to the usual pressures of adolescence and young adulthood, they may be feeling especially lonely, sad, or confused. When your illness is prolonged, the extra time and attention your teens may be spending at home can cause them to feel left out by friends, or left behind by boyfriends or girlfriends. In their desire to keep these relationships unchanged, they may be more willing to do anything to please, which sometimes means yielding to pressures that would normally not shake their values.

Drugs and alcohol tempt adults and teens who are going through tough times. They offer the illusion of a quick fix. The short-term pleasure can blind them to the long-term consequences of this dangerous solution. Teens are especially at risk because of easy access to drugs and alcohol, and a heightened sense of invulnerability to long-term health and social repercussions. When parents are struggling with an illness, they may be too distracted to notice any of the warning signs that their teen is having trouble or has started to use drugs or alcohol.

Another issue that takes on unique overtones in teenagers is that of sexuality. Adolescents and young adults experience increased awareness and exploration of their sexuality. The problem is that normal sexual issues of development might get lost during a parent's prolonged illness. You may feel too overwhelmed by the demands of the illness to bring up such a complex and emotional topic with your teens. Therefore they may worry that they may overburden you if they share their typical concerns, insecurities, or problems related to their sexuality.

Your specific illness may be part of the problem, and thus may become a barrier to talking openly with your teens about intimacy and sexuality. For example, when a parent has a type of cancer that involves sexual or reproductive organs, or when treatment renders the parent infertile, the teens can develop concerns about their own sexuality or fertility. Tension can mount between the teenager and the same-sex parent.

To complicate matters, your teenager's normal sexual issues of identity and limits may become intensified during your illness. For example, if a fifteen-year-old has seen her grandmother die at age sixty of breast cancer, and is watching her mother fight a chronic battle with the same disease, she may feel confused or anxious about her own changing body. While her girlfriends view their budding breasts with excitement as visible signs of their impending womanhood, she fears them as evidence of her risk of cancer. Or feeling her own mortality to a degree uncharacteristic for her age, she may have the urge to do as much grown-up living as possible as soon as possible.

Sexual relations may be pursued as a way of solving problems and feelings that really are not sexual in nature. A teen's identity, ordinarily in flux at this age, can get more confused during your illness. When you make major decisions without their input, they may feel that you are not respecting their opinion or ability to handle important matters, or that they are being neglected. Sex becomes a way to assert their desire to be perceived as mature. Or they may feel burdened by the adult responsibilities they've been asked to assume. Angry or resentful, they may decide to assume the "adult pleasures" too. If your teens feel oppressed by the seriousness or sadness at home, they may be looking for fun or escape. And if they miss the tenderness and security that they always received until their parents became too preoccupied by the illness, your teens may look to peers and other adults for this intimacy.

All of these factors put teens at risk of using sex to solve problems related to your illness such as their anxiety over their own vulnerability, their sense of powerlessness against your disease, or their feelings of abandonment, loneliness, or sadness. In the short run, sex can make teens feel whatever they want to feel: important, grown up, nurtured, safe, loved, happy, powerful, or rebellious.

Addressing sexual issues up front gives you the best chance of protecting your child from encounters that may result in venereal disease, unwanted pregnancy, or unpleasant memories of sexual relations. Promiscuity in teens can be as dangerous as drugs or alcohol. But many parents who aren't dealing with cancer have a hard time talking with their teens about sex; how can you be expected to do it while fighting cancer?

If you've discussed sexuality before your illness, you can just expand on past discussions in the context of your teenagers' overall adjustment to your illness. Listen to your teens. Find out if they are feeling lonely, left out by friends, a diminished sense of self-esteem, or confused about their identity. Help them see that sex, like drugs and alcohol, offers the illusion of a quick fix but can have lifelong negative consequences. Even when things are going smoothly at home, this can be an opportunity to share your concerns and values regarding sexual intimacy, and teach your children about the many ways that people who love each other can express affection. Try not to confuse the normal trials of growing up with the special concerns related to your illness. Helping your children to negotiate their budding sexuality is just part of being a parent.

Choosing to delay or avoid talking about sex is the same as deciding to leave your children to work it out for themselves. Your teens are dealing with these issues whether or not they can talk about them at home. If you can't even think of discussing emotionally charged topics with your teenagers, because you are too sick, too stressed, or just are not communicating well with them right now, ask another adult who enjoys a good relationship with your child, and who shares your philosophy, to do it for you. As with all other aspects of caring for your children, your job is to make sure their needs are addressed, not necessarily to tend to all of them yourself.

Helping Teens Sort Out Conflicting Loyalties

Encourage your teens to continue the work of separating and developing confidence in their independence. In particular, you can provide the encouragement and support they need to deal with the pull of conflicting loyalties. They can have unpleasant feelings no matter what they are doing because they are unsure how to be supportive of you without jeopardizing their social or scholastic life. **Parents can help in a tangible way by offering sound guidance for when the children must be involved with the family, and when it is fine to be off doing their own thing.**

When you discuss your teens' various activities, be sure to separate your needs from your child's when you talk about prioritizing choices. Help your teens do what is best for them overall. For example, let's say that a father is in the hospital getting a bone marrow transplant, and his sixteen-year-old son is ambivalent about continuing on the varsity basketball team because he would have to be at the gym every evening. The teen feels pulled by clashing drives. On the one hand, he wants to play ball and satisfy his needs to vent tension, escape seriousness, be loyal to his social group, and have fun. On the other, he wants to visit his dad in the hospital with the rest of his family, and satisfy his need to be with his father, keep up with what's happening in his family, and feel that he is being a responsible and loving son. The parents can help their child see that both desires are respectable and responsible. After reassuring him that participating on the team is important, his parents can help brainstorm ways to satisfy both needs, such as by visiting his dad at different times, or asking if the coach would stretch the rules and allow him to miss one practice a week, and so on. After presenting all the possible permutations, and discussing which ones are acceptable, the parents should let the child decide.

The next step is very important—**the parents need to support the child's decision**. One child may feel best visiting the dad for a few minutes after school and continuing an uninterrupted schedule with the team. Another child may feel most comfortable sacrificing a season of play in order to be with the family every evening. In both cases, the child must be assured that his or her parents would be fine with either decision, and that both choices are equally responsible and respectable. This approach works if there is a mutual understanding that the parents will let the child know when the circumstances have changed, and therefore when the choices have changed. If the dad has to undergo surgery and feels superstitiously that he won't survive unless his whole family is in the waiting room, the dad's needs take precedence over the teenager's. Or if the father's condition becomes terminal, in most cases the children should be with him even if they don't feel it is their top priority.

For whatever combination of rational and emotional reasons, teenagers may appear to prefer foregoing a chance to say good-bye in order to play in the championship basketball game. Care must be taken to look beyond the immediate, superficial situation. If their dad is lucid and wants to say good-bye, or if the mother needs the emotional support afforded by her son's presence, the parents' needs take precedence. Most important, the parents have an obligation to protect the children's overall well-being by looking at the big picture. Skipping the deathbed scene may be easier on the son right now, but he would lose an irretrievable opportunity to participate with the family in a crucial life event. You don't want your teenagers to find out ten years later that their dad or mom really needed them at the hospital, or that there were people who could have helped them overcome their fears so that they could say good-bye.

There still will be individual kids for whom the best solution in this situation remains to play ball, but it must be a positive choice. All family members must feel that the basketball game is a life-affirming statement of their love. When the family members disagree on what to do, professional counselors such as social workers can help the family arrive at the best solution. Given the immediacy of terminal illness, **investing in counseling maximizes the chance of finding a solution that everyone can live with before the window of opportunity closes**.

As with everything else about survivorship, teenagers must

be informed when the circumstances change, so that the best choices can be pursued, and so that additional support can be provided when needed. Explain to your teen that good solutions for today's problems may no longer work when the circumstances change. Make a pact that you will keep them informed, and let them know when you feel that their decisions may hurt you or them. Ask them to keep you informed of what's happening in their heads, and what they need. Reassure them that you will keep their needs in mind.

When a parent is ill, especially if a parent is critically ill, teens may suffer not only from conflicting loyalties but from a mix of conflicting feelings. They may be flying high with a new love at school, only to come home and feel suffocated by the drawn curtains and the hush surrounding a parent's sickbed. They may alternate from day to day, even hour to hour, from wanting to be treated like a grown-up to yearning to be pampered. Try to be sensitive to what they need at the moment. Reassure them that they are not losing ground in their growth. It may help to say something like "It's OK if sometimes you want to be taken care of like when you were younger. Everyone, even a grown-up, who is in a crisis or under chronic stress has periods of needing extra attention and comfort. If you let people treat you special, you'll feel better and be able to get back to your usual activities. I've seen people who tried to act as if a crisis was no big deal, and got into problems. They got really tired, and angry inside. After a while they had trouble getting their work done, and keeping things together at home."

You can even share with your teen that there are times when you need to be hugged like a little kid, and that when you let someone coddle you, you feel better and can deal with everything more easily. **Home can be your teens' haven if, and only if, they feel that these moments will be kept confidential. Image is critically important to them, and must be preserved.**

Teens and Finances

A practical issue that carries special overtones with teens is that of finances. Financial strain at home is often one of the consequences of obtaining effective and expensive cancer treatments.

Teens are old enough to start understanding the bigger money concerns, and to contribute monies earned at an after-school job. Few teens can contribute in any significant way to the big-buck items—rent or mortgage, hospitalizations, college tuition, and so on—even if they were to quit school and work full time. Your teenagers benefit when buffered from financial stresses that they can't fix. Their responsibility to the family must be balanced against their right to pursue their studies and after-school activities. Open, honest, compassionate exchange will help you and your teen arrive at the best decisions, and support each other.

In summary, the teenage years can be a tumultuous time for parents and kids under the best of circumstances. When a parent has cancer, the stresses can be even greater. However, the insights gained through your illness can put the transition of the teenage years in perspective. Despite the added difficulties, you may end up better able to give your children the roots that ground them and the wings that free them.

Caring for Children in a Single-Parent Home or When the Parents' Marriage Is Unstable

The life-enhancing philosophies and advice that help happily married parents also apply to couples with unstable marriages, and to those individuals who are raising a family on their own. Although applying a piece of advice may require a twist or addition, the underlying principles usually hold true. In some cases, the advice has greater worth because there is less of a safety net for children being raised by one parent, and there are additional tensions for children whose parents don't get along.

Unstable Marriages

When a marriage is on rocky footing, the introduction of cancer can be good or bad. For some, it serves as a catalyst to reevaluate life's priorities, and motivates both parents to work together against the common enemy that threatens their family—cancer. Through the illness, and often with professional counseling, parents learn how to communicate, negotiate, and compromise. In the end, their love is rekindled and the family moves forward after the cancer experience stronger and healthier than it has been in years.

Hopefully, yours will be a marriage that is saved and strengthened by the challenge of cancer. But don't look for a fairy tale transformation brought on by the diagnosis. This rescue is the product of dedication and hard work toward a shared goal, in this case renewing the marriage while learning how to deal with serious illness.

If things have been shaky for a while, or if things are becoming explosive under the strain, get help. In most cases, to say "We'll just get through the illness first, and then we'll work on the marriage" is to put the cart before the horse. Illness is a load on marriages. Marriage counseling, like the horse's harness, offers the possibility of moving forward. If the cart is too heavy for the horse, the cart won't move. But you won't find out if your marriage is beyond repair until you try to mend it. Professional counseling can help give your marriage the best shot.

Unfortunately, a happy ending is not played out in many homes. Cancer is stressful under the best of circumstances. For unstable marriages, cancer can be the final straw (actually, more like a bale of hay). Couples who don't communicate well, whose love is long lost, or who have very poor coping skills often lack the motivation or ability to deal with the extra stresses. Even with professional counseling, the underlying ill-will or lack of affection or respect may be too great an obstacle to overcome in the name of the children.

When separation or divorce is inevitable, the big question is not *if* the parents should split up, but *when*. There are advantages and disadvantages to separating during treatments, and the best timing for each family depends on a host of specific factors. As stated in earlier chapters, **the number one priority is getting the parent with cancer well again**. Many couples destined for divorce make the wise decision to stick it out through the cancer treatments. In these cases, although the relationship lacks love and affection, the parents are able to be civil to each other, and tend to the children. The ill parent is temporarily spared the distractions and strains of divorce proceedings, and is able to focus on treatments and the children. The children are not forced to face the trauma of divorce while coping with their parent's illness. After treatments are completed, there are still the challenges of recovery to face, but both parents can feel that they gave their best shot against the cancer. Now the parents aren't juggling the demands of treatment, and the short-term prognosis is clearer. From a practical point of view, there will be relatively more time and energy to spend helping the children through the divorce.

If, despite both parents trying to keep things smooth, staying together in the same house only causes undue distress for everyone, then sticking it out until treatment is over is not the best choice. Unrelenting marital tension can make it hard for a parent to eat well or rest adequately during cancer treatments. Hateful interchange between parents hurts the vulnerable children by discouraging open communication and by threatening any sense of stability. Sometimes staying together is worse than splitting up.

Divorce and cancer are twin traumas for the adults and their children. Children must be convinced that the marriage had serious fractures before the cancer diagnosis. Otherwise they may develop harmful attitudes about illness, believing that important relationships don't survive. As with all children of divorced parents, they must understand that the final split was not because of them. Remember what has been mentioned repeatedly throughout the text: children tend to blame themselves for things that go wrong. Take advantage of the literature and support services available for children of troubled marriages or divorce. You have to help your children through the fallout of your divorce as well as that related to your illness.

Special Issues of Single-Parent Homes

When there is one parent, children's concerns may be amplified. Learning that a remaining parent is sick stirs up quiescent feelings of sadness and abandonment related to the prior loss of a parent to death or divorce. Unlike children who have not yet lost something precious, these kids know how much loss hurts.

The usual feelings associated with the potential loss of a parent can be exacerbated by memories of the bad times surrounding the loss of the first parent. The myth of invincible parents was shattered with the earlier loss, depriving these children of the benefits of denial. Also, they enter the cancer experience with a repertoire of skills for grieving and coping that may or may not be healthy.

When the sick parent is the sole guardian, there is danger of the children growing up with a debilitating fear of abandonment. This risk can be minimized with appropriate and repeated discussions. First, as discussed in Chapter 5, make clear the distinction between *leaving them*—separating physically because you have no choice—and *abandoning them*—choosing to forsake them. Then follow the steps that help every mother and father allay their children's fear of abandonment: First, make arrangements for your children in the event of your incapacitation or death, and let your children know of the plans. Take care in determining who will raise your children. If your children are old enough, listen to their input regarding your proposed plans and keep an open mind to any concerns or suggestions they may voice.

Taking care of routine matters and your own physical needs can be overwhelming without the support of a dedicated spouse. The benefits of getting help when you are a single parent cannot be overstated. You need your energy for getting well, and for helping your children. Find out who is willing to help, and then call on them. Find out from the hospital social worker or the local office of the American Cancer Society which individuals or groups are available to help out single parents. **Superparent is the stuff of comic books. Seek help and ye shall find.**

Even when you have older children who appear able to tend to their own and your needs, seek outside help. As urged in the previous section on teenagers, remember that older children may have the bodies of adults but they are still children. Your older ones may be able to assume responsibility for their own rooms, and even help out with the meal-making and housekeeping. But taking on too many extra chores, such as car pooling, laundry, or grocery shopping, for a long period of time can be too draining, physically and emotionally, for growing children. And expecting them to assume adult tasks, such as caring for younger siblings, is unwise and unfair for everyone. **The family's needs must be balanced against the personal needs of each child**. How do you judge what is too much? If your children's three basic needs are being met, and they are able to prosper in their academic and social circles, then they are probably OK.

Your children need to know what to do in the event of an emergency. Make contingency plans with one or more backup plans for what to do if you or a child needs to be hospitalized, if you don't show up after school, if the toaster oven catches fire, if the toilet overflows while you are at the doctor's office, and so on. Present these situations as a means of preventing problems and allaying anxiety. Talk through the strategies with your children until they say that they feel comfortable with their responsibilities. Help them practice in their minds what to do in various situations. Try to present all this in an atmosphere of pride and support: "Every kid should be learning these things. I'm so proud of you knowing what to do if . . ."

It's not unusual for mounting medical bills or the parent's inability to work as usual to put families under serious financial strain. Your children need to know that your family will not starve or be thrown out in the street. Even if you spurned financial assistance when you went on your own, you owe it to your children to find out about programs and services available to people in your situation. You're not looking for a handout; you're using services that are intended to help families like yours through hard times.

Your children will benefit from having one or more adults to whom they can turn for comfort and advice. Talk with your kids about who would be a suitable confidant, and whom they would feel comfortable calling if a problem arose. Reconnect them with adult friends, neighbors, spiritual leaders, school counselors, or teachers who could serve as surrogate parents. Reassure them that their connection to others reinforces rather than replaces their relationship to you.

Having survived a divorce or death together, you and your

children probably share a very special relationship. It is natural to want to look to your children for emotional support, but they can be harmed by trying to bear the burden of your emotional distress. You *do* help your children when you let them know if you are upset, and accept their words of comfort and inspiration. And your older children may be mature enough, and willing, to help you work out some of your concerns. But **leaning on children as your sole emotional support can hurt them. Even when you are sick, or feeling vulnerable or needy, you are the *parent*.** Seek outside support from friends, extended family, clergy, and/or counselors no matter how content you feel when with your children. The advice and emotional support of other adults can complement that provided at home, and can serve as backup for you when your teenager's attentions and energies are unavailable.

Caring for the Well Spouse

When a parent has cancer, the well spouse dons many hats. While playing companion, cheerleader, and caregiver to your ill spouse, you have to function as communicator with the outside world, and both mom and dad for various periods of time. Of course, your precancer world did not evaporate upon word of your spouse's diagnosis, and outside demands having nothing to do with illness continue. If you are the family breadwinner, work demands continue unabated, and income may be threatened if you don't fulfill obligations. If you are a homemaker or are independently wealthy, you have more flexibility with your schedule. Still, you will pay a price for sacrificing what you used to do with your time in order to meet the new demands.

The literature that addresses the needs of the caregiver is expanding. Both of you would benefit from reading some of the books suggested in the appendix, and researching resources specific for caregivers. Only by understanding your needs, as well as your role in the care of your children, can you maximize your ability to help your children.

Karen and Mike were high-school sweethearts who married and settled down into a busy routine. Karen chose to be a stay-at-home mom after their second child was born. She kept her outside interests alive by volunteering at the children's school and participating in a volleyball club. A few years later, two months after the birth of their third child, Mike was diagnosed with metastatic colon cancer. The next few weeks were a flurry of activity—getting second opinions and then starting treatment, learning about the disease, taking care of all the insurance and disability paperwork, and tending to the children. As the months passed, the excitement surrounding the crisis faded and they settled into a new routine around his treatments.

Karen's roles were clear-cut, and she tried to fulfill her responsibilities as wife and mother as well as possible. In addition to the usual child care and housework, she provided loving support to her husband when he couldn't sleep because of physical discomforts or anxiety. She was sensitive to her children's usual and special needs. Friends and family were willing and able to help, and Karen welcomed the occasional respite from car pooling and making meals. She appeared to herself and everyone else to have risen to the occasion and to be coping as well as anyone could expect. Unfortunately, many subtle forces were pulling her down. Daily, person after person greeted her with "How's Mike doing? What's happening with Mike?" Not only was she tired of repeating the same news, even when it was good, she hated being reminded about his cancer when she was thinking about something else. On a less conscious level, there was the insidious message that, compared to Mike's situation, what was left of her world didn't rate even passing mention. Since his medical problems took priority over all the things she did to keep the household running, she suppressed the voice that sometimes wanted to cry "What about me? I still have a life, too."

Toward the end of Mike's treatment, Karen came down with a mild case of the flu. Achy and tired, she went to bed in hopes of sleeping it off. Within minutes, their hungry baby started crying. Karen wished that Mike would volunteer to feed the baby. When he didn't, she gently asked if he would. When he complained, "I feel awful. Can you do it?" she got up.

Karen *was* the stronger and healthier one of the two of

them. But she needed rest, too. Before Mike's illness, they would each accommodate the other depending on who needed a break more. For the months of Mike's treatment, the pattern had shifted: whenever she was tired, he was more tired. When she complained of a bad night's sleep, he'd had a worse night's sleep. His problems were always worse than hers.

The stark fact was that Mike's problems *were* worse than hers, and Karen knew it. So she began to withdraw, keeping her difficulties and feelings bottled up. Karen was torn between believing that his needs took priority over hers, and feeling resentful that her needs were never being acknowledged, let alone being met. One day, despite being exhausted, she left home to run an errand and fell asleep at the wheel. Other than a broken arm, she escaped serious injury. The accident jolted family, friends, the health-care team, and Mike into recognizing the strain to which Karen had succumbed. Her problems, although less urgent and not life threatening, were significant. Karen needed and deserved rest, sympathy, and encouragement, too.

Another powerful force that weighed Karen down was the agonizing sense of powerlessness. She couldn't do the most important thing for the most important person in her life: she couldn't fix Mike's cancer. This reality caused simmering sadness and anger that tainted her mood every day. On a rational level, Karen knew that she handled challenges well and couldn't expect herself to perform magic. Emotionally, her confidence was damaged. She thought, "Moms and dads are supposed to make things all better. We put Band-Aids on scraped knees, repair leaky faucets, and work out problems with school teachers." Because she couldn't cure Mike, she became sensitive to other things she couldn't do. When little things went wrong, especially when they were her fault, their significance was blown out of proportion.

I can identify with her anguish. During the months of my first round of chemotherapy, we struggled to keep up with the demands of work, home, and my treatments. On more than one morning, we'd try to get the new day going only to discover that we were out of milk. This meant that Ted had to rush out to the corner convenience store, a hassle that cranked up his internal stress meter. Or at the end of the day, after Ted had showered and got into pajamas, he'd flop down on the couch in the TV room, finally free to relax with me after our kids were asleep, and I'd go in the kitchen to fix my nightly snack of milk and cookies, only to discover the milk problem. Sometimes when this happened, Ted would shift back into work mode, reassure me that it was no big deal, and get the milk without much fuss. But as the months wore on and his reserves dwindled, the little milk problem sometimes triggered an outburst of anger and frustration. That usually upset me, which made him feel bad, and round and round it went. "We're out of milk" became our metaphor for empty emotional and physical reserves and for things feeling out of control. Interestingly, keeping our refrigerator well supplied with milk became one concrete way that I could help Ted. On the bad days, he could say, "At least we have milk in the fridge!"

One aspect of my illness that compounded this problem was that there was no schedule to my good and bad days. Ted couldn't pace himself, or brace himself, because he never knew what to expect. There was some rhythm to the chemotherapy treatments and checkups, but the pattern of my moods and energy was too complex to decipher. Little problems did arise over the course of the months of my chemotherapy. More than once, between the time he left to teach his night class and the time he got home, I'd gone from feeling OK to having a problem that needed medical attention. The few unexpected medical complications that prompted emergency hospitalizations kept him on guard for the next crisis.

Even though most days went smoothly, the day-to-day uncertainty menaced Ted. He couldn't plan important meetings or trips without worrying about possibly having to cancel them because, unexpectedly, I couldn't tend to the children. His insecurity about the short-term future took a terrible toll. For one, he had to pay the consequences, real or imagined, whenever he did have to leave a meeting or cancel one at the last minute.

Knowing that he might be called home, or be faced with a new problem on arriving home, made it harder for him to lose himself in his work. I call this the "rancid jelly bean phenomenon": no matter how many sweet, fresh jelly beans are in a bag, it's hard to enjoy any of them after eating the one or two rancid ones that somehow slipped in among them. So guarded, Ted was deprived of the complete escape and sense of fulfillment that work could have provided.

When friends came over and covered for him when he had professional commitments, he still worried about what was going on at home. If something had happened to me and he wasn't there to take care of it, he'd feel terrible. Not that he could have prevented the problem if he'd been home, and not that our friends couldn't handle the situation as well as he could. It's just that I was his wife, and he felt that he should be the one to be there if something happened.

Another common source of distress is ambivalence about leaving the troubles at home, even when it is appropriate. Ted knew that he needed and deserved a break, and that I wanted him to get away for a while. But he was reluctant to go to the movies when he knew that I couldn't go because I couldn't be in a crowded public place because of low blood counts. When he agreed to go to a party, he often couldn't enjoy himself because of a nagging emptiness associated with being without me. Whenever he became engaged in the conversation, invariably someone would say something, or he would see something, that reminded him of me at home, and suddenly he felt awkward because he was having a good time while I was home in bed.

All of these feelings of ambivalence and guilt are variations of survivor's guilt, a sense of fault for being well. Of course, you are not to blame for being healthy, and nobody wants you to be sick, too. Survivor's guilt is a natural reaction that taps into the big questions of life, such as "Why did this happen to my spouse and not to me?" and "What am I supposed to do with my life since I've been spared?"

Another unpleasant phenomenon related to getting away is that you truly escape for a few hours and, upon walking through the front door of your home, are jerked back again into the whole reality of cancer and seriousness and problems. Crossing back and forth from the well world, the contrasts are sharper and more painful: you are well and your spouse is sick; people "out there" seem to be having fun while you and your spouse are dealing with serious issues all the time; other families have the time and energy to work on their gardens or take vacations while you are just trying to get through each day; other couples talk glibly of future plans while you are uncertain of even having a future together, let alone what trials it may hold in store. All of this can make the well spouse feel any combination of the feelings anger, self-pity, sadness, loneliness, and depression.

An important emotion that may surface is fear of abandonment. Your own past experiences, even childhood traumas, may have made you vulnerable to fear of abandonment. This anxiety may surface when your spouse withdraws appropriately because of physical limitations related to the cancer treatments, or emotional distress. Even if you don't have any special issues of abandonment, you would be expected to feel the loss of attentions that your spouse gave you before he or she fell sick.

While your spouse may be concerned about dying from cancer, you are dealing with the possibility of surviving the loss of your spouse. All of the feelings and issues discussed in books for surviving the loss of a love are compounded when there are children in the equation. The possibility of being solely responsible for the financial, emotional, intellectual, and spiritual stability of your children is daunting. Even when it isn't justified from a medical point of view, your dread of being totally responsible may create a mountain of worries.

In most homes, there is a division of labor. After some time, the parent who normally does not do a certain task feels uncomfortable with the idea of taking it over. If I were the well spouse, I'd worry about being responsible for the bills and taxes and taking care of the yard. My husband worries about how he would guide our two daughters through puberty. We both worry about dealing with other family challenges as the sole parent.

These are a few of the thoughts and emotions that you, as the well spouse, may experience. Obtaining sound knowledge, finding and nourishing hope, and taking action can help you deal with all the concerns that arise during and after your loved one's diagnosis. Learn about the disease and its treatment, especially if your spouse is too sick to do it. The time you invest in this education will repay itself manyfold by preparing you to help your spouse with medical decisions and treatments, recognize problems that need attention, and know when to call the doctors if a problem arises.

Find out who is available to help you with all aspects of daily life, from car pooling the children, to sitting with your spouse when he or she is in the hospital. **Take advantage of the help.** By keeping your energy bank account filled, you will have reserves to deal with the minicrises that pepper every family's week, as well as the major crises that can be part of successful cancer treatment.

Establish a confidant with whom you can share your troubles. Always dealing with your spouse's needs at the expense of your own may leave you vulnerable, the victim of unmet needs and unexpressed emotions. Suppressed anger, disappointment, frustration, sadness, and fear can leak out in ways that are injurious to your family. Fighting and hiding normal emotions is exhausting, leaving you with less energy to tackle things you could handle otherwise. **You can't afford not to spend the time and energy it takes to examine how you feel, and find healthy ways to deal with your emotions.** Always putting your spouse's needs first may sound noble, but it is not realistic or healthy for you or your family.

Find and protect escape time that rejuvenates you and doesn't harm your family. Recognize that time as yours, and allow yourself to forget about the problems, responsibilities, and miseries at home for a while. You are doing your family a favor by making deposits in your emotional energy account. Doing something that makes you laugh or inspires you will help you keep perspective on your situation: there is still joy in the world, you still can feel happiness, and life will not always be like this.

Find the balance between being empathetic and separating yourself from your spouse's pain. You are not helping anyone if whenever your partner feels miserable, physically or emotionally, you do, too. One of the most difficult things for me to accept was the emotional pain my illness caused my husband. It upset me to see Ted worried or sorry for me when I felt awful or had to undergo procedures. Although sometimes I felt jealous when he would exercise or play sports while I was sick, I also was relieved and happy for him. Don't let your ill spouse's understandable moments of envy keep you from enjoying life and doing what you need to do to stay well.

Most important, take care of yourself. Pace yourself. Express yourself in the company of loving, supportive people. Recognize your limits. Forgive yourself when things go badly. Congratulate yourself for jobs well done. The best gift that you can give your family is a healthy you. How to take care of you is discussed in the next chapter.

SUMMARY

Caring for Teenagers

- Maintain house rules and teenagers' responsibilities as much as possible and reasonable.
- Reassure your teens that their thoughts and actions cannot cause your illness; they are not responsible for your recovery.
- Allow respite time for your teens.
- Help your teens stay connected with their friends.
- Be aware of the temptations of drugs, alcohol, and sexual activity to teens who feel stressed or insecure at home; addressing these topics up front in a loving, supportive way can help protect your child.
- Offer your teens guidance for when they must be involved with the family and when it is fine to be pursuing their own interests.
- Your teens' responsibility to your family's financial health must be balanced against their right to pursue their own studies and careers.

Caring for Children of Single Parents and Unstable Marriages

- The parent with cancer needs the best chance at getting well.
- Couples destined for divorce should determine if it is best to proceed immediately or wait until the treatments are completed. Counseling can help couples make the best decision.
- Take advantage of the literature available on divorce.
- In single parent homes, tame your children's fear of abandonment.
- Get help with your chores and emotional concerns from other adults.
- Balance the family's needs against the personal needs of each child.
- Prepare your children for possible problems and emergencies.
- Find suitable adult confidants for each child.

Caring for the Well Spouse

- Read books about or attend a support group for caregivers.
- Meeting the physical and emotional needs of the caregiver is vital to the well-being of every member of the family.
- Regular breaks from seriousness and responsibilities help the caregiver avoid burnout.
- Take advantage of offers of help; ask for help.
- Enjoy life when you can.

7

Taking Care of You

Listen to the sage advice of the flight attendant: "In the event of a drop in cabin pressure, an oxygen mask will automatically drop in front of you. If you are traveling with a child, place your oxygen mask over your face first. Then help your child with his or her mask." These words of advice have much to teach the parent with cancer: **In order to help your children, you must take care of yourself.**

One of your toughest hurdles is finding a healthy balance between taking care of your children and taking care of yourself. Many factors upset the equilibrium in a home where a parent has cancer. For one thing, throughout your treatment and recovery you will have greater physical, emotional, and spiritual needs than before your diagnosis. Many of you will have more needs than you've ever had in your adult life. This does not imply that you aren't handling things well, or that you are weak in any way. All human beings have baseline requirements, and additional ones when they are stressed.

Not surprisingly, your children and spouse, too, have increased needs related to change and uncertainty. In most families, **everyone needs more than usual**. Going through treatments can be tiring, so you and your spouse have less energy left to deal with your children's increased demands and problems. The longer you delay or can't respond adequately to your children, the greater the chance for new problems to develop and small ones to get out of hand.

The first step toward taking care of yourself is avoiding unnecessary work and stress. How do you do that? There is a two word answer: **accept help**. When possible, get practical help with day-to-day chores (meals, laundry, errands, car pooling) so that you can devote your energies to your and your family's needs. You may be thinking, "But it is physically possible for me to make meals." This may be true, but if it's at the expense of missing quiet time with your children or using up your last little bit of energy, then you are doing your entire family a disservice by refusing help. Friends and family who offer to help really want to serve in whatever way helps you best. You need time with your children—time to deal with the painful issues, to escape through fun activities, and to share closeness. Many parents feel guilty asking someone to bring dinner so that they can play miniature golf with their children, but that is what the family may need most.

My children have benefited when I've accepted help and therefore conserved my energies for them. **Finding nuggets of time when I could focus my attention on them, free of distraction or interruption, was worth its weight in gold because they did better.** I could not have created a sense of routine for them, or given them that attention, if I had been physically exhausted or if I were grocery shopping.

One way to conserve energy is to appoint a guard, someone or something to protect your time and energy. Designate a family spokesperson who can keep the rest of the family updated so you don't have to tell the story over and over. Use your answering machine to field calls while you are resting or with your children. Make it clear to others when you need a break from phone calls to preserve your energy and your children's private time with you. You may worry about hurting the feelings of people who care about you. Indeed, some people may be annoyed or angry at your seeming rebuff. Only a superhuman being can take care of everyone and everything. **You owe it to your children to do what is best for your family.**

The value of getting help continues for as long as the extra demands persist. Instead of getting easier with time, for many parents it gets harder and harder to ask for help. They worry about wearing out their friends, or using up their favors. This problem gets complicated as the short-term crisis of a new diagnosis evolves into a treatment routine or series of crises that may last many weeks or months, or even years.

When you are in the hospital, or at home recovering from surgery, you don't have to explain the value of outside help. In contrast, if you are simply tired from monthly treatments, you may look pretty healthy to others. They may not recognize your ongoing need for help, and you may feel embarrassed to accept help when you look fine.

The straightforward approach often works best. Explain what you can and can't do, and what would help you. Spread out the requests for help. The more people you invite to lend a hand, the less of a burden it will be on any one person. I saved myself a lot of needless worry by making a pact with anyone who took an extra turn at carpooling or watched my kids at their house, saying "I'm trusting you to say no if I ever ask for help on a day that's not good for you, or ask you to do something that you'd rather not do. I realize that you have lots going on in your life, too." Then I trusted my friends to keep their end of the pact.

In addition to getting practical help, get emotional support. Work through your feelings of shock, anger, depression, loneliness, fear, anxiety, bitterness, and confusion. Reach out to friends, family, support group, or counseling. It's almost impossible to help your children through their negative or painful feelings before you've dealt with your own. You need to find and nourish life-enhancing philosophies before you can teach them to your children in any meaningful way.

For many survivors, especially those with chronic disease, your feelings will need to be worked through over and over and over. The same issues and problems can resurface throughout survivorship. New cancer-related problems can surface months or years following treatment. People who think that support is only for crises forego an effective way of dealing with the longer-term stresses of surviving. Unchecked, these tensions can destroy your quality of life. Try to take care of all your needs on a regular basis. In sickness and in health, making daily deposits in your energy account provides the reserves needed to tide you over while you tend to your children's immediate needs.

During those times when you are too vulnerable, physically or emotionally, to invest in working through your children's immediate concerns, you may have to address them with a Band-Aid approach. Help your kids find a way to put the problems aside until you can really deal with them, or have someone else help them until you can take over. You're not going to help them very much if you try to deal with their problem(s) when you're exhausted or in a bad way emotionally. Sometimes your attempts may make things worse. **It's OK for your children to see that you can't do everything all the time.**

Despite your best efforts, communication with your children may become stalled or problems may persist. How do you handle problems with your children if you're not sure what's going on or what to do about it? For everybody's sake, get help. Avail yourself of the experience, knowledge, and energy of family, friends, clergy, schoolteachers, professional counselors, and the children's pediatrician. Your whole family is better off if you don't wait until problems develop or get really out of control before you ask for help. Preventing problems helps everyone. Learning to detect them early allows you to intervene when solutions are more readily and easily available.

Sometimes the problem is one of gender. For example, if Mom is in the hospital with breast or ovarian cancer, only another woman may be able to help the pubescent daughter deal with some specific concerns. Other times the problem is generational. Teenagers often feel that the only people who understand them are other teenagers. Or children may perceive their loving, wise grandparents as too old to talk about "stuff we need to talk about with Dad."

As mentioned in earlier sections, find someone with whom each of your children will communicate. The person(s) they choose may not be your first choices, but if the person basically agrees with your philosophies of life, and your children feel comfortable enough to open up, the person is a good choice.

Ensure confidentiality between your children and the person(s) in whom they confide. At the same time continue to nurture whatever lines of communication you have with your child. There's a danger in assuming that the child's nonparent confidant is dealing with everything adequately. **Nothing completely replaces the communication between you and your kids.**

You can learn and trust the best way to address this or that issue under this or that circumstance. Knowing what to do, and doing it in real life, are two different things. I can think of innumerable times when I've been impatient with my children. With embarrassing frequency, they've borne the brunt of my anger or frustration about my illness. In particular, when the need for another biopsy makes me feel my life is spinning out of control again, I can go berserk when my kids mess up the family room. In the setting of cancer, having a tidy living space helps me feel that I can control *something*. The worst part of it all is that I know better. I know that my irritability with the kids is really about the upcoming biopsy. I know that leaving dirty clothing and used dishes in the TV room should not be treated as a felony. As I yell at them, I'm saying to myself, "This is stupid, they don't deserve this. I'm going to be upset with myself later," and so on.

Intelligent, well-intentioned parents make mistakes and handle situations poorly. There are occasions when parents don't have the time, energy, or patience to deal with their children's needs in an ideal way. Medications, fatigue, or emotional stress can cause you to jump on your children unfairly. When you're not proud of how you handled something, tell them you goofed and that you're sorry. Remind them that some days are hard for you. Then forgive yourself. The best you can do is the best you can do.

Living with cancer is an extenuating circumstance. For most of you, this is uncharted territory that stretches your energies at a time when resources are low. As parents, your energies are always being tapped. Whether you were prepared for it or not, **from the moment your children were born, their needs became yours.** When they were hungry, you needed to feed them. When they needed companionship, you needed to be there. Until your children are grown, you have a duty always to balance your personal needs against the needs of your children. Their needs factor into your decisions about what to eat for dinner, whether to take a new job in another city, and whether you should proceed with a bone marrow transplant.

The contrast between the cancer experience of my friends with children and those without has clarified for me what it means when a parent has cancer. Let's look at two of my girlfriends, both my age, each with a type of cancer that responded to intensive treatments. The first friend, Alice, was single, childless, and passionate about her hobby—art history. After her diagnosis, she was able to use all of her energies exploring and pursuing treatment options, developing her spiritual self, and participating in various support groups and private counseling. She traveled whenever possible between her treatments. During these minivacations, she reconnected with friends and family, repaired relationships, and let go of those that were irretrievable. She was doing everything humanly possible to get the best medical care and to tend to all of her physical, emotional, and spiritual needs. While her disease and treatment were a constant drain on her "energy bank account," her daily life was a constant deposit. More than once I heard her say, "I'm getting rid of a lot of the 'garbage' that used to clutter my life. I'm doing things that I would never have taken the time to do before I got sick. I feel good about what I'm learning and how I'm growing."

Alice's physical and emotional energies dwindled during the last few months of treatment. She was sick and tired of being sick and tired. Talking on the phone offered comfort and inspiration. A group of friends and family arranged to take her, in a wheelchair, to the museum of modern art every three days. The visits provided distraction from her discomforts. She would collapse at home for the rest of the day, but awaken in the morning energized until the next outing. She lived as fully as possible within the constraints of her disease during the months of treatment.

My other friend, Elana, also loved art history. Unlike Alice, she had a young son who was struggling with the many separations from his mom necessitated by the treatments. He started having night terrors and before long refused to sleep in his own room. Elana didn't have the luxury of time to talk on the phone for hours or go to several support groups. Whenever she was physically well enough, her son needed her to read to him, eat meals with him, and help him deal with all the emotions little boys have when their mommy is very sick.

Elana was fortunate in having a loving husband and supportive friends and family. They tended to her son on Monday nights so that she could attend a support group. They'd keep him busy when she needed to talk on the phone. Like Alice, she tried to get the best medical care and meet all of her physical, emotional, and spiritual needs. Unlike Alice, she had to balance her needs against her son's. Many times she had to sacrifice the freedom to satisfy herself immediately when she wanted to rest, talk, pray, cry, scream, escape to a movie, or go to a seminar on relaxation. Many, many times she couldn't do what she wanted to for herself because she had to be there, *and wanted to be there,* for her son.

For Elana, when she was struggling with her own discomforts and distress, listening to her son's sing-along tapes didn't begin to provide the distraction that a trip to a museum could have. If anything, her son's favorite activities were so boring, and his whimpering so irritating, that she wanted a break from mothering. Hyperaware of her mortality, she felt guilty for feeling the way she did. Here she was going through all this treatment so that she could raise her son, and yet she didn't want to be with him all the time.

On one occasion, a group of friends took Elana and her son to an art museum. In the first gallery, her son started to whine, "I'm bored." Her friends took over so that Elana could enjoy the art. But the more anyone tried to entertain him or placate him, the louder he complained. The trip was a fiasco, which left Elana feeling that she had to choose between taking care of herself and her son. Her sense of duty led her to the conclusion that she couldn't take care of her own needs at all. In addition to the constant drain of her disease and treatment on her energy bank account, her son represented another constant withdrawal. She felt powerless to make the same kinds of deposits that Alice could.

I watched Alice and Elana as they approached their last months of treatment. The attention and activities that centered around Alice sustained her and distracted her from the discomforts. In contrast, Elana felt increasingly dragged down by all the demands of her illness and family. She harbored a secret wish to be coddled, a notion that couldn't be realized in the face of her sense of responsibility to her son. Her desire to respond to her son's distress was greater than her own physical and emotional pain.

How could Elana tend to her own needs and still be a good parent? With help from a few sessions of out-patient counseling with a social worker, Elana realized that she could make daily deposits in her energy account, like Alice, but in a different way. When possible, she redefined activities and thus transformed her experience. The plots of her son's picture books could never rival that of a good novel, *but the snuggling while reading with her son could.* Board games remained boring, but she was able to enjoy those times with her son when she focused on the happiness reflected in his idle chatter and the wonderful feeling of his little fingers when he handed over the dice. These loving moments together would be her son's childhood memories, which would comfort him for the rest of his life.

When Elana couldn't overcome her impatience with board games, she steered her son toward other activities that he enjoyed—ones that were more interesting for her. She avoided the exhaustion that followed trips to the park by finding ventures that better matched her energy level. For instance, they made up different games for folding the laundry and putting it away: each would have to think of a name that started with the same letter of the alphabet as the article of clothing being folded, or she would time how long it took him to do the job, and so on. Elana's son had fun and her full attention; she had help with a detested task. When her son felt that he'd had some close time with Elana, he was more agreeable to letting her have some time of her own.

Perceived this way, her parenting efforts began to replete her reserves. **Even the sad moments and angry moments, as painful and tiring as they often were, rejuvenated her by allow-**

ing her to escape the patient's role and be a mother. When she comforted her son, or combed his hair, she was not a cancer patient. She was his mom. Many times these moments between mother and son that used to drain her now nourished her. Activities that used to represent withdrawals were now deposits. Elana, like Alice, lived fully within the constraints of her disease while undergoing treatment. Fulfilling her son's needs fulfilled hers. Savor the joyful moments and be nourished. Meet the hard times and be fulfilled.

SUMMARY

- Accept help; ask for help.
- Find time every day to spend with each child alone, free of distraction.
- Get emotional support.
- Accept your imperfections, and forgive yourself your mistakes.
- Try to keep your energy account balanced.
- Arrange respite time from seriousness and your responsibilities.
- Find and nourish courage.
- Find and nourish hope.
- Accept and give love.

8

Caring for the Children When Cancer Recurs or Becomes a Chronic Disease

Fifty years ago, people newly diagnosed with cancer faced what felt like a flip of the coin: either they were going to be cured or they were going to die. Thanks to advances in oncology, the "flip of the coin" metaphor is no longer applicable because a new type of survivor has emerged: patients who enjoy long and fruitful lives even though their cancer is not cured by the first round of therapy. Some of these long-term survivors go through treatment for one or more recurrences before their cancer is cured. Others live with cancer as a chronic disease, one that remits and recurs or that waxes and wanes under the influence of various treatments.

What does it mean for your family if you have chronic cancer? The task of raising children under these circumstances poses unique challenges. In a few important ways, chronic cancer is different than a first-time diagnosis. You and your children are different, too. Understanding these differences, and learning how to overcome the challenges of chronic cancer will help you and your children thrive.

Many of the ideas and pieces of advice that follow are also

found in earlier chapters, such as the idea that **your children need to talk about what's happening to them. They need to share what they are thinking, what they are feeling, what they are worried about, and what they need and want.** You may be wondering why I've repeated the information here. One reason is that if you are dealing with recurrent or chronic cancer you can address your immediate concerns by reading this chapter first. Then you can read the earlier chapters if you want more detailed explanations regarding specific aspects of helping your children. Another reason is that, if you've already read the earlier chapters but it was a long time ago or in the context of a first-time diagnosis, the old messages and advice may sound new, as you now try to integrate chronic illness into your family's life. Basic information bears repeating here because the stakes are high. A family can muddle its way through a temporary crisis and end up okay, if a little worse for wear; a family that isn't coping well with a parent's chronic illness can come undone and the children can get lost. With knowledge and hope, you and your children can deal with your illness in healthy ways.

Facing Recurrence—the Patient

Everything may appear the same if your cancer recurs: the hospital where your biopsy is done, the doctor who delivers the terrible news, the name of your cancer, and even the time of year. Yet, no matter how similar to the circumstances of your original diagnosis, recurrent cancer is not "second verse, same as the first." In certain ways, recurrence can be a more difficult diagnosis to confront. Shocked again by bad news, this time your numbness and sense of disbelief probably won't be as complete or long-lasting. You've known all along that recurrence was a possibility. If anything, this recurrence is the realization of a possibility you've feared since completing your prior treatment. Your intimate familiarity with cancer can make recurrence harder, too. You know from the start how stressful the medical evaluation and treatment decisions can be. Your anxiety about the upcoming months may be reinforced by graphic memories of the ups and downs of your past treatments. Cancer treatment under the best of circumstances involves extra time, energy, and stress.

In addition, a common change in perception of the future can make recurrence more distressing. When you were first diagnosed, you probably thought of your upcoming treatments as having a beginning and an end ("Once I get through my treatment, my family life will return to normal.") In contrast, you see this recurrence as your re-entry into the world of cancer, and you fear it as a move that might prove permanent. Depending on your past medical history, you may still be depleted—physically, emotionally, and financially. Worn-out and discouraged is a tough starting point for any endeavor, let alone raising children while going through cancer.

Lurking silently beneath everyone's overt efforts to help you is the frightening fact that your last treatment failed to cure your cancer. In almost all cases, your prognosis is worse than when you were first diagnosed. Knowing the statistics can make it harder to feel optimistic or hopeful that you'll get through treatment and end up okay. Your sense of uncertainty is magnified as you wonder if this next round of treatment can control your cancer.

The bad news is that your fear, anxiety, fatigue, medical needs, and insecurity about the future can make it more difficult to tend to your children's needs in helpful ways. The good news is that all of these challenges to healthy parenting are ones you can handle and do something about. Before discussing steps you can take, let's set the record straight: this recurrence is not your fault! Don't fall into the trap of second-guessing yourself with useless musings such as "I should have gone with my other treatment option" or "I should have reported my symptom sooner." Don't burden yourself with undeserved self-blame for such things as not visualizing optimally or not nourishing enough hope, as if these measures could have guaranteed a cure. And, don't go blaming others. With all the talk of the mind-body connection, parents are often tempted to blame their recurrence on their children and all the stress at home. Your kids had nothing to do with your cancer coming back. It's not fair to blame them, and it doesn't help you or your family to blame them. Recurrence is something everyone knew could happen even if everyone did everything right.

Take comfort in the many ways that this recurrence is less overwhelming than your first diagnosis. Cancer and its treatment are now familiar territory, and that's good. You already have a working relationship with your oncologist, and a support team that you can call back into action. For the most part, you know what's happening and what you have to do. All of this leaves you more time and energy for your children. It's reassuring to know that instead of starting from scratch you can build on the foundation established earlier for both getting good care and also taking care of your children.

Learning about recurrence helps you help your children, too. For instance, many people assume that they'll need more aggressive and toxic treatments than the first time. Consequently, they are more anxious about their immediate future and may be less attentive or reassuring with their children. Although it is true that many patients need more intense treatment, many other patients do well with treatment that is less toxic. This happens when better therapies have become available since your last round, or when your cancer is the type that is sensitive to less-toxic second-line therapy.

Reject the dangerous myths surrounding recurrence, such as the one that recurrence is always the beginning of the end. That's just plain wrong. **Recurrence is an illness, not a death sentence.** Depending upon what type of cancer you have, your recurrent cancer may be very curable. Even if not curable with today's therapies, as long as your disease is treatable you have reason to hope for a good outcome. For your children's sake, you need to know that this is not the time to give up hope, even if the statistics are unfavorable. Many people have good outcomes in bad situations, and new treatments are becoming available that can change the course of your disease.

My own situation is a case in point. In 1990, the median survival for patients with my type of cancer was seven years. When my cancer recurred barely one year after completing intensive treatment, my prognosis worsened and moved me into a higher-risk group. Over the subsequent few years, none of the cancer treatments I received cured me, but they did keep me alive and well long enough for fludarabine-based chemotherapy and monoclonal antibody therapy to become available. The five courses of treatment I received between 1993 and 1998 weren't even options when I was first diagnosed! It is now 2004, and my cancer is in complete remission, fourteen years after my diagnosis and more than six years since my last treatment. What is my prognosis now? Nobody knows, because the new treatments I've received have opened the door to increasingly longer and stronger remissions. Recurrence is treatable. Keep hoping.

Facing Recurrence—the Well Spouse

If you are the well spouse, here you go again! Ready or not, this recurrence forces you to once again alternate back and forth between the sick and well worlds. Hour to hour, even minute to minute, you have to shift between dealing with the conflicting demands of your spouse's illness, your own work and home responsibilities, your friends and family, and your children. Your life is turned upside down, yet most people are inquiring about your spouse, not you.

No matter how committed a couple you are, dealing with recurrence can pull you two apart, at least for a while, which makes it harder to work together as a team in helping your children. Your spouse is now a patient, wrestling with fears revolving around the upcoming treatments or the possibility of death. You are grappling with very different fears, ones that roared awake after the quiet of remission: fear of losing your life partner, raising your children by yourself, jeopardizing your job or financial security, to name just a few. Fears regarding problems that you couldn't even imagine at the time of the first diagnosis may now occupy center stage, such as the fear of a particular medical complication your spouse could develop, worry that your marriage will crumble under the strain, or concern that your children's mental health or school performance will suffer.

Fear is only one of many unpleasant emotions stirred up by recurrence. Witnessing your loved one's suffering is uniquely painful, cutting to the core in a way that's hard for the person with cancer to appreciate. At the same time, juggling the demands of cancer and normal life with children is maddening ("This is so unfair! My family paid its dues, already"), frustrating ("I can't do anything well because I'm trying to do so much"), alienating ("People don't understand"), lonely ("I don't have time to relax with friends or find out what's going on with other people"), and sad. The depth of your sadness parallels your list of losses that accompany this recurrence. All that you've worked long and hard to regain, like hundreds of dominoes painstakingly lined up side by side, are lost with the tap of two words, "It's back." If you feel discouraged or defeated from the start, you might be wondering, "How can I muster the strength to help my kids through this again?"

Don't despair. You can take steps to deal with these difficulties. As a starting point, remind yourself that this recurrence is not your fault. It's not uncommon to feel guilty, as if you could have or should have done something more to prevent this from happening to your loved one even though you know on a conscious level that you didn't cause the malignant cells to replicate. Don't burden yourself with undeserved guilt that wastes energy. Simply put, cancer doesn't recur because of marital stress, financial troubles, the strain of raising children, or periods of pessimism or poor attitude. When dealing with cancer, nobody can guarantee a cure no matter how much expertise, attention, money, and positive attitude is thrown at the disease. Every time my cancer recurred, my husband found it helpful to steer away from asking "Why?" and to focus on "What now?"

All of the advice I presented in the section on the well spouse dealing with a new diagnosis (pages 119–126) applies ten-fold when you are faced with recurrence. In particular, get information and help. For instance, if you used up your sick leave at work to help your spouse through past treatments, find out about options regarding flexible hours, workload, and so on. If you are in financial trouble, seek the assistance of a financial counselor. Get help if you are feeling torn by conflicting loyalties, such as taking care of your spouse versus tending to your children. Your job is to make sure your family's fundamental needs are being addressed, not necessarily to address them yourself.

At least for a while, you can expect your spouse's needs to be increased. For example, the past few months or years of check-ups and follow-up scans may have become ho-hum. If so, your spouse has been taking care of these doctor visits and diagnostic tests solo with no fanfare at all. With this recurrence, he or she may suddenly want and need a companion, because recurrence is anxiety-provoking and important decisions may need to be made. Remember that you don't necessarily have to be the one to accompany your spouse to every doctor visit, test, or treatment. You just need to ensure that *someone* is there, and that you stay connected emotionally (and by phone) during this tumultuous transition.

Labels affect people's perceptions. Everyone calls you "the well partner," the one who is healthy. The danger is in thinking that this means you can (or should) do everything for everyone all the time. Maybe you tried to be Super Spouse the first time around, and maybe you even pulled it off. Take a moment to reflect: How much did this approach cost you? How much would it cost everyone—you, your kids, your family and friends, your coworkers—if you tried to do it again? It's one thing to push yourself to the limit in a crisis that lasts a few weeks or months; it's quite another to dig deep repeatedly for prolonged periods of time. Ask for and accept help tending to your children's needs. Be sure to let people tend to your needs, too! **When going through chronic cancer, accepting others' support frees you to give what matters most to the people who matter most in your life.**

What if your friends and extended family seem tired of helping? On the one hand, it may be that they would want to help if they realized your family still had needs. They may hold the mistaken impression that you have everything under control since you are a veteran at taking care of your spouse and handling the children. On the other hand, they may recognize your ongoing needs but be unable to help anymore. If this is the case, don't take it personally or interpret it to mean you are asking too much. For all you know, they've been shielding you from their own personal stresses that demand their attention now. You may do best to tap into other resources for assistance. You owe it to your spouse and children to get the help you need to take care of them and yourself.

At each phase of treatment and recovery:

- Learn what is happening to you in the context of this recurrence.
- Find out what you can do to help your family.
- Find out what you can do to help yourself.
- Find the confidence, strength, and support to do what you need to do.

Facing Recurrence—the Children

Your children have been through cancer before. One crucial factor makes dealing with recurrence different than your first diagnosis (or makes this recurrence different than an earlier one): your children are older. Anytime your medical condition takes a shift in direction, your main jobs are

- assessing the impact of this change on each of your children,
- determining what each child needs now and throughout your treatment, and
- ensuring that your children's needs are met.

Not surprisingly, just as you may find the diagnosis of recurrence to be a more difficult adjustment in certain ways, your children may be having a harder time, too. If they were infants the first time around, they were too young to be much affected. They were happy as long as they were well-fed, dry, and loved. Now that they are walking and talking, they are aware that you are hurting and that you are less available to them, two problems that can upset them terribly. Even if your kids were teens when you were first diagnosed, chances are good that they were still too young and inexperienced to project what your cancer diagnosis would mean for them over the subsequent few weeks, let alone months or years. With this recurrence, your children may appreciate immediately the implications of your new diagnosis, and they may be very upset or angry.

In 1990, other than asking if I was going to be okay, none of my children said much after Ted explained why I was suddenly in the hospital. In contrast, in 1994, when I came home from a biopsy and told my kids that my lymphoma had recurred, seven-year-old Jessie started crying and nine-year-old Becky darted out of the house and down the block to her friend's house where, I learned later, she collapsed on their sofa, inconsolable, knowing full well what "It's back!" meant for me and for her.

Becky was upset and angry. "How could this happen *again?* I'm not going through this again!" Some children accuse their parents, "How could you let this happen again?" as if the parents had some control over it. They are putting the blame in the wrong place, but their anger is justified. Through no fault of their own, they have to deal with your cancer again. It's not that these children are selfish and don't care about what you're going through; it's that their world is once again threatened, and they are reacting to something they don't like. Anger is normal and okay as long as it doesn't keep them from doing what they need to do, and as long as it is temporary. As a parent, even if they scream at you ("You are ruining my life!), *don't* accept their accusations. You didn't voluntarily choose to have a recurrence, so don't start blaming yourself for their problems.

Your children's intellectual development affects their reaction to your news, too. Throughout your past treatments, your kids may have been old enough to know you had cancer but too young to appreciate the ramifications. For instance, if your children were four or five, they knew the word "death" and could use it appropriately in a sentence but weren't yet old enough to process its permanence (see Appendix 1). Consequently, they never really worried about you dying even if they knew your disease was life-threatening. If your kids are now five or older, they may be terrified and need help calming their fears.

In addition to the increased needs that arise from their better understanding about your illness, they may find the stress of your recurrence intolerable due to pressures totally unrelated to your illness, such as their concerns about school or friends. And, since they are older, they may feel more responsibility to help you feel better or to help take care of things at home. If they can't help you out in any significant way (or really don't want to), they may feel an ugly mixture of guilt, resentment, helplessness, worthlessness, or anger.

Don't be surprised if your children's emotional needs now are *less* compared to throughout your first diagnosis. This is not uncommon, depending upon their ages, circumstances, past experiences, and personalities. For example, throughout your first course of cancer therapy, your absences may have been a terrible strain on your youngster. You may have spent hours cajoling and hugging your child before and after each of your treatments. Now that he's busy with school and after-school activities, your upcoming absences may not affect him much. Or, while the first time she may have needed frequent reassurances that you weren't dying, the fact that you got through treatment before may be enough for her to let go of this worry, and to see another remission and recovery as the likely outcome.

Since your children's needs change over time, it's only logical that **the best approach for today is one that is in tune with your children's current needs.** One danger is blindly depending on an approach that was successful before. Sure, your previous method of coping may be perfect for your kids once again, but maybe it won't work well now that your children are older. For instance, a child who was proud of your bald head when she was four years old may be aghast of the prospect of you losing your hair when she is six. You do her a service by switching from "If somebody doesn't like my bald head, that's their problem!" to "Let's find creative ways to keep people from noticing I'm bald." As another example, brief daily updates may have been necessary to keep your child comfortable even when everything was status quo, but now, discussing only major changes or milestones regarding your cancer may be the best approach for this same child.

Beware the danger of thinking that you don't need to go over specific territory you covered before. Nothing could be further from the truth. Your children's fund of knowledge, communication skills, emotional maturity, and ability to understand complex ideas or envisage the future has grown significantly since the last time they faced your cancer. This is especially true if your last treatment was completed more than a year ago. In the life of a child, one year is a long, long time. A three-year-old is far more capable than a two-year-old; a sixteen-year-old is often dealing with a completely new set of concerns than a fifteen-year-old. Their needs have evolved, as have their skills and areas of independence.

All of the information and advice proffered in Chapters 1 and 2 about breaking the news applies equally well here, but it is important to note three key differences:

- Since you are breaking the news to children who have experienced cancer before, you need to respect what they already know and refresh them on the prior information that they may have forgotten; finding a balance can be challenging.
- Problems that you've worked through in your own mind (and that, for your sake, would best be left in the periphery) may need to become a central focus of discussion for your children's benefit.
- Topics and episodes that were handled badly in the past, and persistent half-truths or outright lies about your illness, may become obstacles to open communication today.

If it's been a while since you've read the early chapters in this book, you might want to review their concluding summaries in the context of your newly diagnosed recurrence.

From your children's point of view, your recurrence may have the dramatic impact of a first-time diagnosis or the blasé effect of dreary old news. Since it's hard to predict how your children will react or what they'll need, I'll repeat a key message: **reassess their *current* needs in light of your current situation and their current situation, and find an approach that will help them now.**

The Truth Pact

You may be tempted to keep your recurrence a secret. If so, take a moment to think about the effect of sending the following message to your children: "What is happening to me is too terrible to share with you. I don't think you can handle it." Whether you mean to or not, that is what you are telling your kids when you exclude them from the news of your recurrence. This holds true irrespective of your prognosis or how optimistic you are about your future.

In contrast, including them communicates a message of confidence and empowerment: "I'm telling you what's happening because I respect you and have faith in you that you can deal with it." For your two-year-old toddler clinging to your skirt or your nineteen-year-old child off at college, including them is a powerful way to foster lifelong self-confidence. Even more important, sharing the truth is how you establish and maintain your connection so you can comfort and guide them toward healthy ways of coping with the stresses and changes.

It is reasonable to wait until you know something definite or have decided on a treatment plan before sharing the news, but trying to keep your recurrence a secret for more than a few days jeopardizes the bond of trust that is essential for you and your children getting through the stresses and changes as safely and easily as possible. Another advantage of telling them the truth is that doing so puts boundaries around their lively imaginations. Just as before, your children know that something major has happened even if you don't say a word. If anything, their antennae are raised to detect little changes that other kids might miss. Your children can pick up on the most subtle of signals—a few extra phone calls, a slight edginess to your voice, or a parking receipt from the garage at your doctor's office lying on the dashboard of your car.

Formally re-establish open lines of communication about your illness and about whatever else is going on in their lives. Personally, I think it helps to be repetitive and overly simplistic. I usually say something like, "I want to go over a few things about what's going on. You are helping me when you listen for a few minutes. Even if you think what I'm saying is old and boring, I feel better when I'm sure that you know and understand these things. And, kids, you are helping me by keeping me up-to-date on what's going on with you. I want to hear about your schoolwork and friends, and all the stuff that has nothing to do with cancer." This approach serves three functions:

- It gives your children a tangible way to feel that they are helping the situation.
- It emphasizes the necessity of open communication.
- It reinforces that they are supposed to go on with their lives no matter what is going on with you.

What convinces me that this method works well for my children is how often they respond to old information as if it were new, and how they listen with interest when I'm reviewing something I'm sure they already know. Believe me, this is definitely not the case when talking with them about most anything else!

Each time my cancer recurred, my kids and I ceremoniously renewed our pact, no matter how corny it appeared. I would actually cross my heart with my index finger and say, "I promise always to be honest and keep you informed, and you must promise to do the same." They might roll their eyes or sigh, and that's fine with me as long as they also acknowledge commitment. Then, I'd remind them that no matter what is happening with me, unless I tell them otherwise, their main job is to continue being a kid. I expect them to do their schoolwork, sports, music, and other after-school activities. I expect them to stay involved with their friends. I can't tell you how many times this approach has helped us deal with stresses and problems unrelated to cancer, too. Life is hard. Life's problems are easier to deal with when my family doesn't keep secrets. Nothing is as comforting and inspiring to me or my kids as knowing we are in this life together.

Your Children's Needs This Time Around

As always, tell the truth, couched in love and hopefulness. State clearly that recurrence is an illness, not a death sentence and that you and your physicians are working together to get you better. Go over the essentials again: cancer is not contagious, treatment may make you sick before it gets you better, and it may take a long time to get you well.

Compared to dealing with a first diagnosis, the danger today is that you'll assume they already know what they need to know. If you are going to assume anything, assume they don't remember from your last bout of treatment, or assume that they remember but still don't understand completely. Repeat. Repeat. Repeat until they tell you, "Enough, already! I know!"

You might want to ask each of your children what he or she remembers about cancer, in general, and your cancer, in particular. You might be surprised as much by what they don't know as by the facts and images that remain vivid in their mind. Going over the basics about your past and current condition is a good way to identify and clear up any lingering misconceptions that may have slipped by unnoticed. I've found that misunderstandings are easier to catch if I ask an open-ended question like, "Tell me what you know about why I'm going to lose my hair," than if I rattle off, "Chemo causes hair loss."

Explain in plain language what is planned regarding your evaluation and treatment. If you don't yet know, reassure your kids that everything that needs to be done is being done. Matter-of-factly tell them that you'll all have to live with the uncertainty—and *can* live with the uncertainty—and that you will inform them of the details as soon as you know anything.

In addition to sharing the news, **emphasize that this is not their fault.** It doesn't matter how many times you've told them that they can't cause your cancer. It never hurts to say once again, "In no way is this recurrence your fault. Nothing you've ever said, done, or thought caused this to happen." Even when they know this to be true, it helps to hear it again and again.

If your children appear at all distressed, validate that their situation is hard while focusing on the positive sides of dealing with recurrence as compared to when you faced cancer before. For one, since your family got through it before, you *know* something now in a way you could only *hope* before: "We *can* get through it." You also have the advantage of knowing what approaches worked well for your family (and what ones didn't) when it came to dealing with cancer-related stress, anxiety, waiting for test results, bad moods, hospitalizations or other separations, and juggling treatment with everything else that happens in a busy household.

For your kids' sake, mention things that are still right in their world: The seas did not freeze, the rivers did not stop flowing, they still have food on their table and a roof over their heads, their favorite television shows still air in their assigned time slots, and most important of all, your love for one another is as strong as ever.

Your children need a safe place to work through their hard times and painful emotions. Children need a place to vent how angry, sad, unhappy, disappointed, frustrated, lonely, and frightened they feel without someone trying to cheer them up or encourage them to "be positive!" They often need guidance and support exploring or working through their feelings so they can move forward. If they are unhappy, validate their misery while also giving them encouragement that they won't always feel this way. Talk with them, or connect them with an adult who can help them find a way through their unpleasant feelings. Without a doubt, seeing my children's emotional distress was hard on me. Letting them express themselves honestly was easier for me when I kept my goal in mind—doing whatever I had to do to help my children thrive. Mind you, it was still hard.

Take note: your children may remain totally unruffled by the news of your recurrence, as if nothing important has changed. Don't assume that they are in denial or putting up an optimistic front. Many kids adopt what is called "healthy avoidance" as a coping tool in a difficult situation. Being able to somewhat disconnect from an unpleasant situation is a common and highly successful method for kids to deal with circumstances over which they have little control. Not getting too emotionally involved saves their energies for the normal tasks of growing up—schoolwork, friends, and so on. And, some kids simply don't feel anxious or perturbed by a parent's recurrence because they feel confident everyone will get through it and things will work out. *As long as your kids know and understand what's going on,* they benefit when their genuine optimism is validated, not squashed. Give them a place to share their confidence and lack of worry. Ideally, the safest place to speak their mind will be your home. If you can't absorb either their painful emotions or carefree attitude right now, that's understandable. Your job then becomes finding someone who can be there for your children—your chaplain; a close family member or friend; or a social worker, child psychologist, or cancer counselor—until you can do it yourself.

Living with Chronic Cancer

When talking about helping your children, the similarities between chronic and recurrent cancer outweigh the differences. In general, chronic cancer is when the cancer does not go into a complete remission, therapy is ongoing, or no available therapies are likely to lead to a durable remission or cure. With no end in sight, you have to live with doctor visits, treatments, follow-up tests, decision-making, and the ups and downs that accompany illness. For many of you, physical limitations and the emotional stress will affect how you care for your children. For practical purposes, whether you are dealing with a recurrence that will take a long time to treat, or you are dealing with chronic cancer, your concern is how best to help your children when cancer is going to be a part of your life for a long time.

The biggest difference between chronic cancer and a new diagnosis is adjusting to the possibility of living with cancer indefinitely as opposed to finding ways to just get through it. If you screw up while dealing with a one-time diagnosis, you can pick up the pieces and still move forward. If you don't find healthy ways of dealing with chronic cancer, your children may spend the remainder of their childhood in a home marked by neglect and unresolved problems.

Another difference may be your self-confidence and sense of determination. It is one thing to hang in there with a time-limited challenge, and quite another to deal with difficulties that seem to have no end in sight. As an analogy, if you are told that you need to lift a heavy weight for as long as possible, you may give up after grunting and sweating for thirty or forty seconds. If everything is kept the same but you are told that you only need to lift the weight for one minute, your expectation of stopping soon gives you the fortitude to endure to the end.

It is a major adjustment to realize and accept that your disease has become chronic. In contrast to the discrete moment of diagnosis when you suddenly became a cancer patient, the shift from recurrent (or persistent) cancer to chronic cancer is a period that is often blurry. Like a rising tide, new waves of fear, anxiety, anger, sadness, and confusion may wash over you for a while as you adjust. As you settle into new routines, you may feel uncertain about how best to integrate your cancer into your life. How much time and attention does cancer have to get? When can you let your illness drift into the background, as if it didn't exist? You may also have times of wondering if problems with your kids are due to your illness or are simply normal age-related phases. Another challenge is not feeling sorry for yourself or your kids, and learning to balance hope and expectation when dealing with chronic disease.

Lastly, long-term survivorship can be accompanied by ongoing losses that affect your children in profound ways. Losses may be chronic, such as when families are dealing with a parent who has become oxygen-dependent, wheelchair-dependent, hard-of-hearing, or otherwise affected by cancer treatments. As the children grow up, they may experience the consequences of their parent's longstanding loss differently over time. For example, a toddler may adjust well to a parent's inability to walk, and years later may experience the loss anew when she realizes that her parent can't accompany her on amusement park rides or chaperone a camp-out. Or, the cancer-related loss may first occur years after the parent is cured of cancer, such as when a parent's chronic problem progresses, or when a parent develops a late effect of treatment. As an example, a child may adjust well to her parent's inability to run or play sports due to mild lung injury from chemotherapy. Years later, the child may face new losses associated with her parent needing oxygen all the time due to progressive lung dysfunction. Grief is not a one-time job but a gradual process that, in children especially, can come in fits and spurts. **Your children need permission to grieve old losses that they are experiencing in new ways, as well as permission to grieve new cancer-related losses.** The expectation is that grieving will help them adjust and move on.

Creating a "New Normal"

When cancer recurs or persists, dealing with cancer becomes part of your family's normal. For you, **the challenge is creating a healthy sense of normal under exceptional circumstances.** For your children, the challenge is living with the knowledge that their "normal" is not a typical "normal." Potential dangers for your children include

- developing a self-perception as unlucky,
- fostering habitual self-pity,
- nurturing unrealistic expectations of others or themselves,
- feeling alienated from their friends,
- feeling disoriented or confused,
- feeling as if they are in a chronic state of crisis,
- perceiving any little thing that goes wrong as a crisis or chaos.

It may help to steer your children away from comparing your family or your situation to those of others, and to point out that every family has its unique combination of troubles and strengths, hard times and easy times, good luck and bad. Reassure your children that you will use your family's strengths to deal with your challenges.

Creating a healthy "new normal" while dealing with repeated courses of cancer treatment was a constant balancing act for me. I acknowledged the many cancer-related problems and stresses we faced: "I know it's harder for you to concentrate in school when I'm in the hospital. I know it is stressful for you to arrange your schedule around my treatments and daily naps. I know you don't like it when Dad and I are tense the week of my treatments."

I acknowledged the extra strain while telling them that I expected them to adjust and deal with it. This attitude encouraged them to move forward in difficult times. I believe that expecting my kids to do their best helped keep things feeling normal. I often told my kids, "I expect you to pay attention in school as best you can. I expect you to do your homework like all the kids whose moms are healthy. *If you can't, we need to find a way to get you back on track.*"

Make an effort to ask your kids about their friends, after-school activities, and social events. Broadening the conversation helps keep your cancer from crowding out these important spheres of their identities, and helps life feel normal at home. It takes a conscious effort to keep illness-related problems from defining the children, but it can be done.

Problems: Cancer-Related or Other-Related?

All parenting involves times of heightened emotions as your kids go through phases, make mistakes, and struggle through disappointments and challenges. Having a family member undergoing cancer treatment can lead to heightened emotions, increased emotional needs, and decreased reserves under the very best of circumstances. When you're having behavioral or emotional problems with your kids, your dilemma becomes dealing with your child's moodiness or belligerence that may reflect an age-related phase or a response to some problem with a classmate. Or, your child's unacceptable behavior may, in fact, be completely due to the stress of your illness.

As I see it, I don't need to determine first if a problem with one of my kids is triggered by or exacerbated by my illness. I need to deal with the problem! If it is because of the stress of my illness that my child is having nightmares or slacking off on homework, the problem doesn't suddenly become more or less pressing. Similarly, ascertaining that a new problem is one that is common at this age doesn't give me permission to relax and let it go. The bottom line is that if we have a serious problem, we need to define the problem and work toward solutions.

Solving family problems—any problems—seems to take on more urgency when I am sick because we don't have the energy needed to handle normal, age-related problems. By addressing tensions and not blowing them off as "a phase" even when they are "a phase," we are more responsive to our children's needs unrelated to my illness. I suspect this makes our house more peaceful and joyful than it would be otherwise. Certainly, when I am in remission and the stress due to my illness is minimal, dealing with problems as they occur helps keep our home a haven.

Choosing Your Battles

I am not suggesting that you deal with every single problem right away. Sometimes the best course of action is to ignore a problem, or to decide to deal with it later. In my home, I'll address immediately any child's sudden use of foul language or dip in grades, no matter what. However, if the tension is high because of a rough patch with my cancer treatments, I'll usually postpone mentioning my frustration with their messy rooms or arguing over which TV shows they watch. Certainly, you can choose not to worry about repainting the shabby living room or resolving some longstanding issues with a mother-in-law. Regain control by recognizing that you have choices. In general, it helps decrease the overall stress level if you resolve important family problems as swiftly and completely as possible, especially problems that are likely to persist or recur.

Avoiding Pity

Pity is poison to children. Sure, it's sad that your kids have to miss out of certain activities and comforts that they would have enjoyed had you not been sick. So, too, some kids enjoy being smarter, faster, stronger, or more musical than yours. That's life. Instead of feeling sorry for your children, try to see the difficulties in your family as opportunities for your children to gain confidence in their ability to overcome obstacles and adjust to unavoidable losses.

Throughout the 1990s, my need for treatment disrupted many a spring break and summer holiday. As it turned out, my cancer often recurred in the early summer, which made it impossible to travel to Canada for the grand annual family reunions. In addition, my chronic problems with limited energy and headaches meant that I couldn't drive long distances or participate in activities that were too demanding or lasted more than a couple of hours. From my children's perspective, I've always been needier and less available than the other moms.

With an aim toward putting a positive spin on my physical limits, I've kept the following in mind:

- This is the way it is, no matter how anyone feels about it!
- Pitying myself or my kids only makes a bad situation worse.
- We *can* deal with this and find creative ways to work around my limits.
- This is nobody's fault, so nobody should be blamed.

When I feel like my illness is making life hard for Becky, Jessie, and William, and when I'm tempted to pity and spoil them, I remind myself that as long as their fundamental needs are being addressed, good things can come out of the hard times:

- When my children work around my limits, they are learning flexibility.
- When my children respect my special needs, they are learning tolerance.
- When my children tend to my needs, they are learning compassion.
- When my children wait for me when I'm slow, they are learning patience.
- When my children see me as whole, they are learning about self-love.

- When my children pick up the slack, they are learning about teamwork.
- When my children fend for themselves (when they are perfectly capable of doing so), they are learning self-sufficiency.
- When my children deal with my setbacks, they are learning perseverance and resilience.
- When my children continue to grow and develop, they are learning that they are worthy, independent individuals.
- When my children nourish dreams, they are learning about hope.
- When my children have times of happiness, they are learning about joy.
- When my children feel loved each and every day, they are learning about love.

During the first few months of my illness, I deserved an Emmy for acting as if I didn't pity them when, in fact, I really did. I countered my urge to coddle them by consciously saying to myself, "I am not going to pity them. I won't let myself feel sorry for them." After a while, I came to believe what I knew intellectually—that pitying them hurts them. Now, I truly don't feel sorry for them when their plans get cancelled or they are inconvenienced by my limits. That's life and they can handle it.

Keep separate in your mind the difference between "not pitying them" and "not tending to their needs." The "no pity" approach can backfire if kids get the message to "buck up and get over it" when they are feeling disappointed, angry, frustrated, or sad, instead of being given a safe place to express what they are feeling, and work through their feelings. When kids are having a tough time, granting them some latitude and offering them comfort and support may be in order. This is not pity but a healthy response to their increased needs as long as the expectation is shared that they will eventually adjust and move on.

I've found that **linking losses with gains helps me and my children.** I remember the day I slept through pick-up time for

William. My kindergartner was left standing all alone at the school curb for at least ten minutes. When I finally arrived, I immediately and wholeheartedly apologized. Then I pointed out for him the positives, by saying, "William, I'm *proud* of you waiting all by yourself and not panicking. And, *you are so nice* to understand and forgive me for goofing up. I think you can feel good about yourself knowing that you can handle it if someone is late." Over the next few weeks, I purposefully mentioned the good sides of that bad situation. In this particular case, the *overall* impact of that bad experience was positive.

I've made a conscious effort to articulate the lessons for them and applaud their successes to be sure the messages are ingrained deeply in their developing minds. For example, "Thanks for working quietly on your homework while I napped. You were generous to give me the gift of a good rest. I'm proud of how you took care of yourself for a while." Or, "I know this isn't the spring break you hoped for, but I'm proud of how flexible you are being. It's good to know that you can have a fun time even when plans don't work out the way you expected or wanted. Being flexible will help you when you are older: flexibility will help you find ways of being happy when things aren't going the way you planned."

Through the challenges of chronic cancer, you can teach your children skills and values that can help them through tough times and find joy. You can show your children their own strengths and teach them the power of love.

Expecting vs. Hoping

If things are not going well right now, or if the outlook is not good, can you tell your children the truth and yet instill hopefulness? Absolutely, yes! One approach is to distinguish between your expectations and your hopes, find a balance of acceptance and hope that works well for you, and then guide your children toward a healthy balance of acceptance and hope that works well for them.

When tackling the hope-acceptance conundrum with your children:

- Explain the essential facts about your illness that have nothing to do with expectation or hope. ("I still have cancer [or I have it again]. I am receiving new treatments that are controlling [or slowing down] the cancer.")
- Explain that you are receiving treatments with the expectation of overall improvement. ("I'm getting this treatment because most people do better with it than without it.")
- Explain in general terms what usually happens to most other patients in your situation, and that this is what your doctors expect to happen with you. ("In other patients this treatment usually slows the cancer for a while.")
- Explain in general terms the good outcome that is *possible* and what you hope for. ("Occasionally patients get a long remission, or even a cure, from this treatment. I'm hoping I am one of the people who do really well.")
- Acknowledge lost hopes and encourage new hopes. ("The old treatment disappointed us and didn't get rid of the cancer like we'd hoped, but now we can hope that this new treatment gets me better.")

Acceptance does not have to crush hopes. Although my family does not expect me to be cured and or return to my precancer energtic self, we benefit from maintaining the global hope that one day I will be cured and my energy will improve. This hope floats in the background while we nourish many hopes that are more short-term, circumscribed, and likely to be satisified. When we hope for "a good check-up" as opposed to "a cure," we know the answer within months, and we are more likely to have our hope satisfied. When we make shorter-range hopes such as the hope that I have a good day, we have more control over our hope being fulfilled.

Help your children see that they can nourish a variety of hopes that revolve around goals unrelated to cancer. For instance, your family members can hope to maximize cooperation and maintain closeness, succeed in school and work, enjoy friends, keep the house reasonably tidy, enjoy healthy meals, and to learn new skills, have new experiences, and make new friends. Talk with your children about your hopes and dreams, and encourage them to talk about theirs. When specific hopes don't come true, encourage your children to let them go and then find and nourish new hopes. Acceptance and hope can free you to find healthy ways to deal with the uncertainty and adjust to the limitations so you can move forward.

Rituals and Routines

Family routines and rituals help create and maintain a sense of normal in the setting of illness. Preserving old rituals and developing new ones gives a sense of wholeness and continuity to your family life no matter what's going on with the sick parent, especially if illness is chronic. As an example, our family relishes bagels and cream cheese for lunch every Sunday. Before we clean up, we have our special once-a-week dessert. Ted watches as I meticulously slice a large black-and-white cookie into eight equal wedges, and the kids and I execute an elaborate routine (with rules too complicated to bother explaining here!) of choosing pieces until all of them are gone. This sweet weekly family tradition is comforting, especially when my illness makes our life feel out of control.

Holiday rituals mark the seasons and can bring families together in a spirit of love, hopefulness, and joy. Celebrating holidays helps your children hold on to their sense of moving forward throughout the year, and over the course of many years. On a more mundane level, develop routines for daily tasks and activities, such as who washes the dinner dishes, where each child goes after school, and what hours the children are allowed to watch television. Routines conserve everyone's energy by eliminating squabbling and everyday decision-making.

Transitions, by definition, involve change, loss, and stress. Celebrations of major milestones serve a useful function by helping you through transitional markers such as birthdays, graduations, anniversaries, promotions, and retirements. In a similar way, creating treatment-related traditions can help your kids deal with the changes and losses surrounding the transitions of your illness. Rituals can help children appreciate a sense of moving forward through hard times. For example, think about throwing a "hat party" before starting another course of chemotherapy. Hang a big funky calendar on the refrigerator and have your kids mark off the days of treatment. Write a jingle for the kids to chant before your treatments or when times are tough, such as "Roses are red, violets are blue, cancer is weak, and the survivor is YOU!" Watch a movie as a family on the Saturday after the beginning of each cycle of treatment. Make a family donation to a cancer-related organization at the end of each treatment cycle. Take your kids to celebrations of survivorship sponsored by your local cancer organizations so that they can feel proud of your (and their) survivorship in a fun and upbeat setting. Be creative! Enjoy inventing family rituals. Family rituals can lessen your children's distress, bolster their hope, dissipate loneliness, and bring joy through the ups and downs of your illness.

Finding Joy

"This is the day. . . . We shall rejoice and be glad in it." Psalms 119:24 is a call to life. Today is your children's childhood whether you are in treatment or not, whether your prognosis is good or poor, and whether you are feeling fit or lousy. For your own benefit as well as your children's, make an effort to find and create a few happy moments every day.

- Savor your children's daily development. Ordinary growth and maturation become extra-special when you appreciate it in ways that families that have never faced illness often don't.
- Celebrate successes, both big and little, both yours and theirs.
- Exercise your humor, and encourage your children to develop their own senses of humor. Laughter helps strengthen your bonds with your children while lightening everyone's load and keeping up spirits, especially when the strain and losses are likely to be long-term.

- Look for the positive outcomes of bad situations, the so-called silver linings.

Every family enjoys its own unique brand of humor. Let humor provide an escape-hatch from the potential oppressiveness of chronic illness. **Just because having cancer is sad, you don't have to feel sad all the time.**

Preventing Burnout

Living with a parent's recurrent or chronic illness puts children at risk of losing interest in life and energy for living. What happens is that, over time, the cumulative effects of stress lead to a child's emotional exhaustion. He no longer feels anything and has no energy for learning, enjoying, or responding to what's happening around him. Like a burned-out light bulb, this child can no longer function.

"Burnout" is a term often used when someone can no longer function well in one specific sphere of life. For example, someone who is experiencing "job burnout" loses energy or enthusiasm for work, but may be feeling and functioning fine outside of work. In this section, I'm using "burnout" to refer to the extreme end of a global process in children who are dealing with chronic family illness, and these children *are not fine.*

What follows are some suggestions of measures that can help families prevent or interrupt the development of burnout in children. For families that are coping well, these measures may help you through the long haul of recurrence and chronic cancer in healthy ways.

Enforce rules. Keep routines. Your specific rules and routines will change as your children grow, of course, as well as when you face the ups, downs, and shifting demands of treatments. Even so, clear-cut boundaries and a general sense of routine are comforting and grounding for your children. If you have times when you don't have the physical or emotional stamina to maintain house rules and routines, ask other adults to act in your behalf.

Create "cancer holidays." As discussed in earlier chapters, your children benefit from regular escape from the seriousness of cancer. This notion takes on heightened importance in the setting of chronic cancer. After all, if cancer is the main focus of your family life for 4, 6, or even 12 months of a one-time diagnosis, your family can still reorient itself after completion of treatment and move on. If cancer is the main focus of your family life when dealing with chronic cancer, your children's whole childhood may revolve around cancer.

On a regular basis, arrange to have your children spend a few days, or even just a few hours, in an environment where cancer is not on the radar screen. Encourage them to forget about cancer when they are in school, participating in after-school activities, or playing with their friends. During relatively calm times, I've made a conscious effort to avoid mentioning anything having to do with illness, especially when it really doesn't add to the conversation or if we are doing something fun. If my kids happen to bring it up in conversation, or need to talk about it, that's fine and I encourage it; but I don't bring up cancer when it really isn't necessary. This way, we avoid the trap of seeing everything as, somehow, related to my cancer.

Maintain realistic and healthy expectations. What do you expect of each child? Is it reasonable and healthy under the circumstances? **Beware the danger of expecting too much.** Requiring your children to take on too much adult-type responsibility or to become very involved in the care of the ill parent can lead to their burnout, especially if your children become focused on home to the exclusion of other areas of their lives, such as school or friends. High expectations regarding school performance or keeping a stiff upper lip in crises can lead to the children's burnout if they feel they can't measure up to others' expectations. At the same time, **beware the danger of expecting too little.** If every failure in school, with friends, or at home is excused with something like, "It's okay, under the circumstances," you are inadvertently teaching your children that failure is an acceptable response to tough situations. You are also sending the message that success and happy endings are possible only when life is otherwise hunky-dory. For most people, life is a matter of succeeding despite obstacles, and finding and creating happiness despite the stresses, little failures, disappointments, and things that aren't going your way.

Make sure someone is listening to your children. When possible, listen to your kids with the kind of focus that guarantees that you would ace a detailed quiz on what each child said. When your children tell you stories about their school, friends, homework, a movie, or anything else that is important to them, put everything else out of your mind as much as possible. Unless you are driving or doing some other activity which demands visual vigilance, look at your child and communicate with your whole body that you care about what he or she is saying. Just five to ten minutes of intense wholehearted listening can make a huge difference for your kids. All those "in-between moments" are prime opportunities to listen, such as while carpooling to and from school or sports practice, while clearing off the dinner table, or while waiting for a dentist appointment. Feeling heard and understood helps immunize your kids against burnout.

Help your children find workable solutions to their problems. Burnout also arises when children have ongoing problems that are not being addressed. Find out if your children are having sleeping difficulties, problems with a friend or teacher, or anxiety about what's happening to you when they aren't home. For all you know, a child's acting out may be unrelated to your current treatment but due to an embarrassing episode at school or a falling-out with a friend. Once you know what's going on, you can guide your child toward satisfactory solutions, or help your child find ways of adjusting to situations that can't be changed.

What if you are doing what you can to prevent burnout, yet you remain concerned that your child is at risk? Symptoms of impending burnout run the gamut: from trying to control everything and everyone to acting uncharacteristically passive and apathetic, from displaying frequent emotional outbursts to withdrawing into a mute state, from becoming fidgety and unable to sit still to becoming sluggish and resistant to any physical activity. Your child's grades can go up or down, depending on if the child becomes obsessive about schoolwork, finds it impossible to concentrate, or sees schoolwork as irrelevant and worthless. Children experiencing burnout have been known to overeat, stop eating, sleep all the time, suffer insomnia, or develop headaches, stomachaches, fatigue, itching, or strange symptoms.

A child's symptoms may change over time if early burnout progresses and is not detected or addressed. At the most severe stage, a child simply cannot function—he or she can't learn, complete normal age-related tasks, or enjoy usually pleasurable activities and people. A red flag should go up if your children develop persistent symptoms that keep them from continuing to grow and develop, or that keep them from enjoying life. As with most problems, the earlier you diagnose and respond, the better the chance for a smooth, speedy, and complete recovery. If you suspect burnout, get professional help. As a start, call your pediatrician, the hospital's oncology social worker, or a child-life specialist.

Recurrent and chronic cancer affects everyone in the family. Actively preventing and responding to problems at home helps keep your children on track and helps them meet the challenges of survivorship. With love and hope, you can teach your children the skills and values that can help them grow strong.

SUMMARY

- Recurrence is different than a first-time diagnosis.
- Your children know that something has happened, even if you don't say anything.
- Tell the truth couched in love, support, and hopefulness.
- Make it clear that recurrence is not their fault.
- Reassess your children's current needs in light of your current situation and theirs.
- Find an approach that meets your children's current needs.
- Acknowledge the challenges while nurturing confidence in dealing with them.

- Provide a safe and supportive place for your children to express their feelings.
- Deal with problems while they are small.
- Avoid pity.
- Help your children nurture acceptance with hope.
- Ask for and accept help for as long as you have needs.
- Preserve and develop family rituals and routines.
- Seek out and create joyful moments.
- Create cancer holidays.
- Enforce rules.
- Maintain realistic and healthy expectations.
- Make sure a caring adult is listening to your children.
- Help your children find workable solutions to their stresses and problems.
- Learn how to recognize and respond to burnout in healthy ways.
- Create a "new normal" that integrates your illness into a home that strives for health, happiness, and hopefulness.

Conclusion

Thirteen years ago, while in my first remission, I began to write about family life when a parent has cancer. Since then, I've been through seven more courses of cancer treatment, and I've raised my children to their current ages of fifteen, seventeen, and nineteen. My oldest has left home for college. Today, thankfully, I am in my sixth year of complete remission. Having never regained the stamina needed to resume clinical medicine, I've embraced the life of a writer. In the mornings, I craft words for articles, books, and lectures. In the afternoons, I rest and then tend to my kids. Although life is not what I had wanted, prepared for, or expected before my cancer, life is good.

During the years of writing the first edition of this book, occasional glimpses of clarity flashed through my brain, such as when I felt the faint breath of death at each of my "bad news" checkups, and when I witnessed the struggles of other families after the death of a parent. These flashes of insight outlined an image of what it means to raise children when a parent is seriously ill. But my image, for the most part, developed as a slow focus through the lens of living with my chronic cancer.

In the years since the first edition was published, my ideas and beliefs have been validated as I've experienced the longer-term effect of lessons learned through hard times. My hope has become reality as, over and over, I've watched my children deal in healthy ways with everyday life stresses, losses, changes, and challenges. My children demonstrate a love of life even though they know and seem to accept that life can be unfair, disappointing, frustrating, and sometimes very painful. I've seen them find hope when the chips are down and create new hopes when old ones become impossible. Instead of fighting or hiding their unpleasant emotions, I've seen them use their emotions to get help when needed, and to move forward always. From what they tell me and what I've seen, their heightened awareness of the fragility of life has caused them to live not in fear but with gratitude. At school and at home, with friends and family, they don't sweat the small stuff, and they savor everyday pleasures in a way I find unusual in young people. Like wizened elders, they seem to "get it." As a parent, I revel in seeing this perspective which arose out of family illness making their lives easier and more joyful.

My understanding of life after cancer is based on more than the experiences of one mom (me) in one particular situation (mine). Through my advocacy work, public lectures, and workshops, I've heard hundreds of stories from mothers and fathers and aunts and uncles and grandparents and teachers and family friends. They've described for me the difficulties and blessings that have accompanied their efforts to raise children when a parent is sick. I've watched as parents chose to keep their secret from their children, and then faced the troubles that followed. I've seen parents and children put up brave fronts throughout their courses of therapy only to fall apart during the recovery period after treatment, or even many years later when the pressure is off and everything is going well. And, I've witnessed family after family enjoy happiness *despite* illness and then tap into the special happiness that can occur *because of* illness: the family lore about courage, ingenuity, and endurance repeated often and with pride and joy; the indisputable, indestructible, ineffable family bond that combats loneliness and breeds confidence in the most difficult of times; and a depth and strength of love that is heavenly.

My current perspective is one of a middle-aged physician, long-term survivor, and mother of three older teens. When I think about the situation of raising children when a parent has cancer, my vision is one of mothers and fathers trying to raise their children when their physical and emotional reserves are depleted. They wrestle with reassuring themselves and their children in the face of uncertainty. They squirm while trying to explain painful truths, and ache when they see their children sad, angry, or frightened. The well spouses struggle to meet the needs of each family member. Most painful of all, the parents with cancer suffer anxiety over possibly being unable to finish their most important life work—raising their children. This anxiety breeds another—a sense of urgency to do it right. Right now.

The children are hurting, too. The complicated task of growing up becomes more complex as some of the innocence of childhood is lost. The ups and downs of the parent's survival challenge any sense of routine and constancy at home, two forces that buffer the normal pains of childhood and adolescence.

What are parents to do? In the face of perceived threat, our natural instincts drive us to protect our children. The reality is that we cannot protect our children from all the change, loss, pain, or stress of life. But we can prevent many of the short- and long-term detrimental effects by meeting our children's physical and emotional needs in an environment of truth, unconditional love, and trust, no matter what the circumstances. The best gift we can give our children is not protection from the world, but the confidence and tools to cope and grow with all that life has to offer them.

You are reading this book because you love your children. This love is your greatest strength. Use the power of your love to help your children heal and grow. In helping them, you'll help yourself. And remember—whatever your prognosis, you can be a good parent. All it takes is courage, hope, and a lot of love.

APPENDIX 1

Major Stages of Growth and Development

Children go through fairly predictable stages of growth and development. Knowing what to expect at each age allows you to be alert to problems, feel relaxed about expected behaviors, and respond in helpful ways. Being able to recognize when your children are in the midst of a developmental transition is important because at those times they may have less ability to tolerate the stresses and frustrations associated with your illness.

Various outlines of the major stages of childhood growth and development have been described by specialists. What follows is a brief overview of the progressive stages of a child's psychological and emotional development, based on a composite of Freud's, Erikson's, and Piaget's theories. The italicized sections highlight more specific factors that may be involved as your children adapt to your illness.

Your child's development may seem to fall outside the described stages. Don't panic. Although this may indicate the possibility of a problem, it may be perfectly fine for your child. Many normal, healthy children do not follow the described stages for one or more aspects of their development. Your pediatrician can help determine if further evaluation is indicated.

Each child is unique. Everyone is best served when adults provide information, comfort, and support in a style that is best suited to the specific nature of the child. As discussed more fully in the text, when you are unable to meet one or more of your children's needs, find appropriate substitute caregivers such as family friends or relatives.

Birth to One Year

During this first year, babies learn to trust and depend on others. They become able to allow their mothers out of their sight without rage or fear, trusting that they will return. Infants focus on their mouth as their primary organ of gratification and begin to coordinate information obtained from the various senses.

Your infant needs predictable contact with recognizable, loving caregivers who tend to their basic need for food, clothing, shelter, and love. Whenever possible, provide some calm and uninterrupted time for you and your infant.

One Year to Three Years

During this stage, children begin to develop their sense of autonomy as they become "potty trained" and learn to recognize the difference between "I" and "You." Stubbornness often reflects attempts to separate themselves from others. Their moods and cooperativeness may fluctuate as they adjust to new situations and build confidence. Although children may begin to use symbols to represent things, their world is understood based on what can be seen directly, with these perceptions dominating judgements.

Your child needs instruction and support to continue the work of gaining control over bodily functions and to feel more comfortable asserting his or her independence. If these developmental milestones are delayed, there is an increased risk of the child having feelings of shame, doubt, fear, and poor self-esteem.

Since your very young ones interpret things based on what they see, and they don't yet have a grasp of time, facts about your illness or treatments may be impossible to explain, and fears difficult to assuage. Their tendency to put more weight on what they see than what they are told may continue to affect their comfort level and ability to adjust, even as their reasoning abilities mature. For example, until they are in their teens, they may find it difficult to believe that your treatments are making you better if you appear sicker than you did before.

Three Years to Five Years

Guilt and conscience are developing. Your children may experience "magical thinking," whereby they see themselves as the center of the universe: Everything that happens is related to or caused by them. Normal, healthy ways to tame anxiety include play, fantasy, and pride in new skills. They further develop their ability to use symbols but continue to draw conclusions about their world based on what can be seen directly or by what they can understand.

Help your children to communicate what they are feeling. Reassure them in various ways that they are not responsible for your illness or recovery. Encourage them in their learning. Nourish their social interactions with others. Let them use play and fantasy to calm anxieties.

Five Years to Twelve Years

These years focus on learning from others as well as by self-exploration. Friendships begin to develop outside of the family, with "best friend" relationships and same-sex peer groups becoming an increasingly important part of their world. Healthy self-esteem develops as they build on their fund of knowledge and repertoire of skills and enjoy success with relationships outside of the family. Without these, children are at risk of feeling inadequate or inferior.

Your child benefits from opportunities and encouragement to explore the world and develop friendships outside the home. When you need treatment that is prolonged or away from home, keep in mind that their conceptions of time and space are still developing. If your child seems to see everything in all-or-nothing terms, try to encourage more flexible ways of thinking. For example, help them understand that cancer may or may not be fatal and that toxic cancer treatments can be life-saving.

Thirteen Years to Adult

Teenagers work towards independence from their families and search for their own identities. Their bodies undergo sexual development, and their body image changes. As they sort through the difference between how they appear to others and their own self-image and as they clarify their self-identity and social role, they often display unpredictable and varying mood swings, ambitions, desires, and levels of dependence or independence.

During these years children develop a scientific understanding of cause and effect and an ability to use reasoning. Their fund of knowledge and repertoire of skills continue to expand, as does their creativity and ways of thinking about the world around them.

Your teens benefit from encouragement and support as they continue to gain knowledge and skills, explore self-identity, and integrate into the world outside of the family. Those teens who want detailed explanations about your illness do better when this information is provided to their satisfaction. For those who don't need to know the particulars, forcing information on them can be harmful.

Teens need the opportunity and support to develop healthy friendships or love relationships. Difficulties may arise when they feel conflict between the demands at home and the desire to be the same as their peers and when all family decisions seem to be governed by the demands of your illness. Acknowledge their conflicts, and, when possible, give your teenagers choices. Help them to maintain a sense of control over aspects of their environment, a prinicple that applies to children of all ages.

Teens try on adult behaviors and can assume some adult responsibili-

ties for short periods of time. When parents are stressed by serious or chronic illness the distinctions between teenagers and adults may become blurred. Teenage children are still children. Don't ask them to be surrogate parents to younger children or surrogate partners for the well spouse. Share your burdens with other adults.

APPENDIX 2

Glossary for Kids

Below is a glossary of cancer-related medical terms, explained in kid-friendly language. Use it as a resource when trying to answer your children's questions or explain medical terms. Definitions are accompanied by analogies, examples, and suggestions for demonstration techniques that may help you clarify meanings. I discourage you from trying to teach children everything in the list, unless they ask to learn it all.

Your children may get frustrated when they can't grasp what you are trying to say despite your many heroic attempts. If you appear calm, it will be easier to reassure them that they don't need to understand everything right away. Let them know that this information is complicated and is hard for you to learn, too. Often they just need more time before the definitions begin to make sense.

Your using the words in conversation will help them, just as it does when learning a foreign language. Be aware that your children may overhear adults talking who are misusing medical terms, or speaking in more ominous tones than necessary. In addition, your children may use a word appropriately, and appear to have a good understanding of its meaning, when in fact their definition may be a little off, or even completely wrong.

Anemia (a-neé-mee-a)—Someone who has anemia has too few red blood cells. Anemia makes a person feel weak and tired. You cannot catch anemia. Anemia can develop if a person doesn't have enough iron, in which case iron pills will fix it. Bleeding also causes anemia. Many cancer patients who have anemia are not helped by iron pills because their bodies can't make new blood cells very well. If the anemia gets bad, the doctors can fix it by giving medicine or blood to the person with anemia.

Biopsy (bye′-op-see)—A biopsy is when doctors remove a little piece of the body where they think there might be cancer. Sometimes doctors can get the piece with a needle that they stick through the skin. Other times they have to do a small operation to get the piece out of the body. Doctors take the little piece that they think might be cancer, and look at it through a machine called a microscope.

Cut out twenty 2-inch squares of white paper and five 2-inch squares of colored paper. Now take five empty cups. Put five white squares in four of the cups. Put five red squares into the fifth cup. Draw an outline of a person on a big piece of paper, or lie down and put the cups on various parts of your body. Ask your children to "biopsy" each of the body parts by sticking their hand in the cup, getting some of the squares, and then looking at what they got. Depending upon how sophisticated you want and need to get, you can adapt this game to explain why your doctors need to repeat a biopsy (they didn't get enough tissue, or they got tissue with no cancer but they are worried that they missed the part with cancer, and so on).

Blood counts—Blood looks red when we cut ourselves, but it is really a mixture of things with different shapes and colors floating in salty water. Some of these different things are called red cells, white cells, and platelets.

Blood counts tell us the number of red cells, the number of white cells, and the number of platelets in a certain amount of blood.

Low blood counts—There are not as many shapes and colors in the blood as normal.

Normal blood counts—The blood is normal.

High blood counts—Some people say "high" when the blood counts are *too* high, but some people say "high" when the blood counts are normal. Make sure that you and your children understand what you mean when you refer to high counts.

Take a tall clear glass or tube and a bag of colored macaroni. In a bowl, mix up lots of red macaroni, a few pieces of white macaroni, and a few pieces of another color macaroni. You can use marbles or colored Legos, or whatever else you have available. Call the red item the red blood cells, the white or yellow item the white cells, and the remaining items the platelets. Put the mixture into the glass or tube. Show the children how, from a distance, the tube looks red even though it's a mixture of different colors. That's why blood looks red even though it's a mixture of different colors. There are many more red cells than white cells in the blood.

You can use this model to demonstrate anemia by taking out some of the red macaroni; low white counts by removing the yellow macaroni; and so on.

Draw a line 12 inches long. Draw a second line, parallel to the first, 1 inch apart. These two lines represent a vein. Now draw small red dots between the lines, about ¼–⅛ inch in diameter. For every twenty or thirty dots, draw one white dot the same diameter as the red dots, and one green dot much smaller than the red dots. Fill in the whole vein with dots. Look at the picture up close and you will see all the different colors and sizes. Look at the picture from far away, and it will look like a red vein.

Bone marrow—The red stuff in the middle of bones is called the bone marrow. This is where blood is made. Then the blood goes out of the bone marrow and travels in the veins and arteries to the rest of the body. The marrow is like a car factory, only instead of making cars it makes blood cells.

Take a bone, such as a big chicken bone. Break it in half and show the children the dark red stuff in the middle of the bone.

Bone marrow transplant—There is a kind of cancer treatment where the doctors give a lot more medicine to kill the cancer than they give with regular (standard) chemotherapy. The medicine is so strong that it also kills the healthy blood-making cells in the bone marrow. So, after the medicine kills the bad cancer and the good marrow, the doctors give the person new bone marrow cells. They do this by dripping the new marrow into the sick person's blood through an IV. After a while, the new cells find their new home in the empty bone marrow and begin to make new, healthy blood.

If the new bone marrow is from someone who is healthy and has never had cancer, the transplant is called an allogeneic (al″-o-je-ne′ik) or allogenic (al″-o-jen′-ic) transplant. The person who gives the healthy bone marrow is called a donor.

Sometimes the new marrow is taken from the person with cancer. To do this, the doctors first remove the marrow. Then they give the person medicines to get rid of cancer in the body, and they also put medicines on the marrow that they removed. This way, they kill all the cancer in the marrow before giving it back to the person. This type of bone marrow transplant is called an autologous (aw-tol′-o-gus) transplant.

Cancer—Cancer is a disease where a tiny part of the body doesn't work right anymore, and it grows too fast and too much. The growing cancer hurts the healthy part of the person's body. You cannot catch cancer from someone else.

Check out some of the books in the bibliography (Appendix 4) for analogies that may help explain the concept of cancer cells crowding out normal cells, or stimulate ideas of your own (e.g., good bubbles, Pac-men, or other creatures that multiply and die in an orderly fashion, versus the sick bubbles, or whatever, that multiply quickly and crowd out or destroy the good bubbles).

Carcinoma (car′-sin-oh′-ma)—A carcinoma is a cancer that begins in the lining of an organ. It is a type of cancer, in the same way a poodle is a type of dog, roller-skating is a type of sport, and a shirt is a type of clothing. Use these analogies to help you explain how all carcinomas are similar (they are all cancer; they all cause trouble if they are not controlled; they can all be treated) yet everyone's cancer is unique (your poodle may look like someone else's, but you can tell your poodle from every other poodle in the world).

CAT scan—A CAT scan, also called a CT scan, is a picture of the inside of your body. CAT scans have nothing to do with cats or kittens. A regular camera takes pictures of the outside of your body. A CAT scanner takes pictures through the skin. It does not hurt to have a CAT scan done, but since the person has to hold real still during the CAT scan, some people feel stiff and tired afterwards.

Cells (sells)—Cells are tiny parts of the body, so small that you can't see them with your eyes, that work together to make the whole body. If you made a big castle with Legos or building blocks, each piece would be like one cell, and the castle would be like the whole person.

Checkup—A visit to the doctor to make sure that treatments are working, or to make sure that everything is still OK after treatments are over, is called a checkup.

Chemotherapy (kee-moh′-ther-a-pee)—This is special medicine whose job is to get rid of cancer. Some chemotherapy comes in pills or juice and can be taken by mouth. Some chemotherapy has to be put into the blood through the hand or arm or through a tube in the chest. The nickname for chemotherapy is "chemo."

Clinic (klin′-ick)—A clinic is a special place where doctors and nurses help people get well and feel better, and the people go home again after they are treated. At a cancer clinic, all the people they take care of have cancer.

Cured—When the cancer is all gone and is expected to stay away for the rest of the person's life, the person is said to be cured. Sometimes the doctors have to wait until the cancer is gone for a few years before they can use the word "cured."

Diagnosis (dye-ahg-no′-sis)—A diagnosis gives the name of a person's sickness.

> Here is an idea to help older children understand the concept of making a diagnosis. Play a game similar to Twenty Questions, where you or your child have to figure out "the diagnosis." For example, your child can pretend to have a broken bone, strep throat, or poison ivy. You can ask questions like "Does it itch?" (no) "Does it cause vomiting?" (no) "Does it hurt?" (yes) "Are you sick all over or just in one place?" (just in one place) and so on. When you can name the medical problem, you have made the diagnosis and won the game.
>
> With younger kids, it may work better if you stay away from medical diagnoses. For instance, you can play a game where the child tries to figure out what is in a paper bag. Use an apple or a stuffed animal. Let them ask questions. Let them touch it through the paper bag while you make the comparison to your doctors feeling a lump under your skin but being unsure of what it is. When your kids have figured out the identity of the hidden object, they've made the diagnosis.

Drug—A drug is any liquid or pill that you take into your body that changes the way the body works. Drugs can be good or bad depending on how they are used. Doctors give drugs to people with cancer to help get rid of the cancer and to help the person feel better.

Point out the difference between drugs used for healing, and street drugs. There is such a big push to sensitize kids to the dangers of drugs with the "Say No to Drugs" campaign that some children may be frightened or confused when you refer to therapeutic drugs.

Fungus (fun′-guss) (the "guss" rhymes with fuss); **fungi (fun′-ji)** (the plural form of fungus)—A fungus is a type of living thing that is like a plant. A mushroom is a fungus. The yeast used to make bread rise is another type of fungus. Most types of fungus do not cause illness, but some can cause sickness if they get into the body. Sometimes a person has to be sick already from a disease like cancer or diabetes before the fungus can make them sick, too.

Hematocrit (hee-mat′-oh-krit)—This is one of the numbers on a blood test result that tells the doctor how much of the blood is red cells.

Hemoglobin (hee′-mow-glow-bin)—Hemoglobin is the stuff in red blood cells that makes them look red. Like hematocrit, this is also one of the numbers on a blood test result that tells the doctor how much of the blood is red cells. It is a different measurement than hematocrit, but the two numbers are related.

Hormone (hor′-moan)—A hormone is a chemical made by the body to help it work right. The body makes many different kinds of hormones. Hormones can also be made by scientists and given like a medicine. Sometimes people with cancer need hormones because the body isn't making the right amount of hormones by itself. Or they may need extra hormones to help them get better.

Immune system—The parts of the body that fight infections and also fight cancer are called the immune system. It may help to visualize the immune system as an army inside of you that gets rid of sick cells.

Induction therapy (in-duck′-shun ther′-ah-pee)—The first treatments given to try to get rid of cancer are called induction therapy.

Infection (in-feck′-shun)—When a tiny living thing that is not supposed to be in the body, like a virus, gets into the body, grows, and causes trouble, the trouble is called an infection. The infection can be caused by a virus, bacteria, or a fungus. The infection can be in one place or all over the body.

Often people refer to infections as "bugs," as in "I've caught a bug" or "I've got a bug." Make it clear that you are not talking about cockroaches and spiders.

Leukemia (lew-kee′-mee-ah)—This is a type of cancer that starts in the blood.

> See the demonstration described for **Blood counts**. Keep adding white marbles or macaroni to show how the white cells are outnumbering and overcrowding the red cells. You may want to use a different color than white to distinguish the normal white blood cells from cancer cells.

Localized (low′-kal-ized)—Cancer that is only in one spot and no place else is said to be localized. In a similar way, an infected toenail is a localized infection, and chicken pox is not localized because it is all over the body.

Lumbar puncture (lum′-bar punk′-cher)—See **Spinal tap**.

Lymph nodes (limf nodes)—Everyone has hundreds of little bean-shaped bumps called lymph nodes in their bodies. These bumps work hard to fight infection and cancer. For some people, cancer can go to these lymph nodes and make the lymph nodes bigger. Finding big lymph nodes helps the doctor know where the cancer is.

Lymphoma (lim-foe′mah)—This is a type of cancer that starts in a lymph node.

Maintenance therapy (mane′-ten-ants ther′-ah-pee)—After treatments have gotten rid of as much cancer as they can, maintenance therapy is given in hopes of preventing the cancer from coming back if it is gone, or keeping it from getting worse. Some types of cancer need maintenance therapy. Other types of cancer don't, because the person does just as well without it.

Mastectomy (mast-eck′-tow-mee)—An operation to remove a person's breast (a lot of people call a breast a boob) is called a mastectomy. This helps the doctors get rid of the breast cancer and prevents it from coming back in the same breast.

Metastasis (met-ass′-tah-sis); metastases (met-ass′-tah-seez) (plural form of metastasis); also called a "met" (plural "mets")—A metastasis is a piece of cancer that broke off from where the cancer started and traveled to a new place in the body where it is growing.

There are innumerable analogies to try. For example, seeds that fall off a tree are like metastases: they are carried in the wind to a distant place where they settle in the ground and sprout into a new tree. Or tell the story of two rabbits on a big island. The rabbits have baby bunnies who leave and settle in a nearby area. After a while, they are old enough to have their own baby bunnies, who move away. Before you know it, there are bunnies living all over the island.

Metastasize (met-ass′-tah-size)—When a piece of cancer breaks off and travels to another part of the body, it is metastasizing.

Nodule (nod-ewel)—This is another word that means lump or bump. A nodule can be cancer, or a cyst, or an infection, or a lump of fat.

Oncologist (on-kah′-lah-gist)—A doctor who treats only people with cancer is an oncologist.

Oncology (on-kah′-lah-gee)—Oncology is the study of cancer.

Operation (op-er-a′-shun)—When doctors make a cut in the skin to fix something underneath, they are performing an operation. Sometimes people are given medicine to make the part of the body having the operation fall asleep. Then they can't feel any pain when the doctor fixes that part of the body. Sometimes people are given medicine to make them go into a deep sleep during the operation. Then the person can't feel anything at all during the operation. After the doctors fix the problem, they sew the cut together again with stitches. When the cut is all healed, it leaves a scar that you can see on the skin. A scar does not hurt.

Oxygen (ox′-ih-jen)—Oxygen is a gas that people need to breathe to stay alive. We breathe air, which has oxygen in it. Sometimes when a person's lungs or heart isn't working well, they need extra oxygen, so they breathe special air from a machine that has more oxygen in it than the air around us.

Platelet (plate′-let)—A platelet is something that is in the blood to help you stop bleeding when you get cut (see **Blood counts**).

Prognosis (prog-no-sis)—The doctors' best guess at how well a sick person will do is called a prognosis. The prognosis gives you some idea of how serious the illness is; it never tells you what is definitely going to happen. No matter what the chances are for getting better, nobody knows for sure what is going to happen. Lots of people who have been told that their prognosis is bad end up doing fine. And sometimes people with a good prognosis get better but then get sick again, or don't get better at all.

Radiation therapy (ray-dee-aye′-shun ther′-ah-pee) (also called radiation)—Radiation is one type of treatment to get rid of cancer. It is done by aiming a special machine at the cancer that hits it with cancer-killing rays. It does not hurt to get radiation, but it does make you tired.

Recurrence (ree-ker′-anse)—Sometimes someone is treated for cancer and the cancer goes away completely for awhile but then comes back. When this happens, the cancer that returns is called a recurrence.

Remission (ree-mish′-on)—After finishing treatment for cancer, if the cancer is completely gone, the person is in complete remission. After finishing treatment for cancer, if there is still some cancer left but it is at least 50 percent gone, the person is in partial remission. Remission is not the same as cure. Remission can last a long time or a short time.

Sarcoma (sar-koh′-mah)—This is a type of cancer that starts in a bone, nerve, muscle, or blood vessel (as opposed to an organ like the liver or lung).

Spinal tap (spi′-nal tap)—This is a procedure in which doctors use a needle to get some fluid from around the nerves in the middle of the back, to find out if there is infection or cancer there. If the doctors find cancer in the fluid, they can sometimes place medicine into the fluid through another spinal tap.

Spleen—The spleen is a part of the body that is inside the belly on the left side, just under the bottom of the ribs. The job of the spleen is to help keep the blood working right and help fight infections. Sometimes cancer can go to the spleen.

Surgery—See **Operation**.

Transfusion (trans-few′-zhun)—When you take blood from healthy people, put it into a special bag, and then drip the blood from the bag into a sick person who needs blood to get better, you are giving a transfusion.

Tumor (tew′-mer)—A lump or bump may be a tumor. It may or may not be cancer.

Viruses (vy′-russes)—Viruses are very tiny living things that are not plants or animals or fungi. They are a kind of germ. They can only stay alive if they are living in other living things, like people. Most viruses do not cause illness. Some cause minor illnesses like colds or warts. Some cause terrible illnesses like polio. Some viruses that don't cause bad sickness in healthy people can make people really sick if the people are already sick from cancer.

X-rays—These are special beams from a machine that can pass through the skin and take pictures of what the body looks like on the inside. The picture is also called an X-ray.

APPENDIX 3

Resources for Parents and Children

Take a few minutes to find out about informational and support resources. If you feel pressed for time, or don't have the emotional energy to deal with this, ask a friend or family member to make the calls. There are relatively few places to call, so I advise calling or writing to most or all of the national organizations and all of the local ones. This way you get a complete picture of what is available to you. There will be some overlap of advice or referrals, but this has advantages. Getting the same information from different sources can help you decide what will work for you. The main reason to invest the time in calling each of the local organizations is that each group may mention a service that is not discussed by the others. Knowing what is available and at what price allows you to compare and find the best combination for you.

In response to the growing demand and diminishing funds for these services, you may find your call answered by a machine. If you can leave a message, do so. If you find yourself listening to endless music, or announcements like "Your call is important to us but all of our operators are busy. Please stay on the line . . . ," you may find it more efficient to send a fax or letter requesting a call (or specific information).

Some of the national organizations can provide information or telephone support directly. Most refer you to local chapters or resources. In addition, organizations for your specific diagnosis, such as Y-ME (breast cancer) or the Leukemia and Lymphoma Society of America, may have additional services geared to advising and supporting you in your efforts to help your children.

National Cancer Institute (NCI)
Building 31, Room 10A16
9000 Rockville Pike
Bethesda, MD 20892
Phone: 800-4-CANCER
Fax: 301-402-5874

NCI has a toll-free number for its Cancer Information Service (CIS) answered by trained volunteers who are prepared to deal with questions about cancer or to direct you to the person who can give the answers. Informational booklets are sent free of charge.

American Cancer Society (ACS) (national headquarters)
1599 Clifton Road, NE
Atlanta, GA 30329
Phone: 404-320-3333

Canadian Cancer Society (national headquarters)
10 Alcorn Ave., Suite 200
Toronto, Ontario
Canada, M4V 3B1
Phone: 416-961-7223
Fax: 416-961-4189

ACS is a nationwide community-based voluntary organization with national and local offices. You can obtain the telephone number of the local office from your oncologist's office or the yellow pages. ACS has trained volunteers, many of whom have dealt with cancer themselves, to answer questions, and provides informational videos and pamphlets. In addition, ACS offers personal and group support. Some cities offer a workshop or a support group specifically for children whose parent has cancer (see below).

The Canadian Cancer Society (CCS) offers many of the same services as the ACS. Provincial divisions are listed in the telephone directory.

American Association for Marriage and Family Therapy
112 South Alfred Street
Alexandria, VA 22314-3061
Phone: 703-838-9808
Fax: 703-838-9805
www.aamft.org (Click on "Therapistlocator.net" on their home page for referral to a local medical family therapist.)

Medical family therapy is a specialty that deals exclusively with helping families function and stay together during or after a health crisis.

These therapists may work through a doctor's office, a hospital, or in a private practice setting.

The National Family Caregivers Association
10400 Connecticut Avenue, #500
Kensington, MD 20895-3944
Phone: 800-896-3650
Fax: 301-942-2302
info@nfcacares.org

This is the only national organization that serves family caregivers, their friends, and the professionals supporting them. Their wonderful quarterly newsletter, "Take Care," offers practical information and emotional support for the caregiver in a format that can reach those who don't have the time, energy, or inclination to attend a support group.

Well Spouse Foundation
63 West Main Street
Suite H, Freehold, NJ 07728
800-838-0879
info@wellspouse.org

This organization focuses on the spouses and partners of the chronically ill and/or disabled. The national office can refer you to their local chapters and support groups. As a member, you can take advantage of their chapter meetings, round-robin letter chains, outreach via telephone and computer, and their free quarterly newsletter, *Mainstay.* The Well Spouse Foundation holds annual national conferences and offers regional respite weekends.

Social Workers

Many social workers are experts in family and child-life issues who are available for individual or group consultation or counseling. They may be a good resource to help you deal with the expected medical, emotional, or social difficulties of having cancer. They can help you identify problems with your children and find effective approaches to resolving these problems. In addition, they can reassure you when your children's behavior is normal and adaptive under the circumstances. You can find a social worker who might be well matched to your particular situation by calling any of the following places.

Local Hospitals

The department of oncology and the department of social services of your local hospital are usually well informed about all the social workers and cancer support groups available in your community. Find out if there are support groups available that are geared to parents with cancer or other serious illness, and to the well children. Groups geared to children whose sibling is seriously ill can be of value, too, since many of the well child's issues are the same whether the patient is the parent or a sibling.

Hospice

Hospice is a program designed to address the physical, spiritual, social, and economic needs of terminally ill patients and their families. If the parent is under hospice care, take advantage of the expertise available for dealing with the children's issues. If the parent does not need hospice care but is expected to in the next year or so, you can call the hospice office and inquire about information or services that would be useful before hospice care is needed.

Churches and Synagogues

Social services are often available through local religious organizations. If you belong to a congregation, call the office and inquire about services available through your church or synagogue and through related community services.

Office of Your Oncologist

Some oncology offices provide information or support services.

Workshops and Support Groups for Well Children

Let me make special mention of programs around the nation that are geared to the special needs of children whose parent has cancer. Programs in your area can be an invaluable resource. All of these programs provide a supportive environment where children can express and explore feelings, learn about the changes caused by illness in the family, and develop effective coping strategies.

Many parents in treatment feel they don't have time to add "one more thing." Those parents who have completed treatment often feel they want to put it all behind them. They don't want to go back to the hospital setting any more than is absolutely necessary. Participation in one of these programs may prove to be your wisest use of time if you or your children are having a hard time adjusting to the illness, the treatments, or the recovery process.

To find out about support groups in your area for children, call your oncologist's office, all of your local hospitals (not just the hospital in which you receive treatment), the local chapter of the America Cancer Society, the local chapter of a specific cancer organization (such as the Leukemia and Lymphoma Society of America, Y-Me, Us-Too), the Wellness Community (www.wellnesscommunity.org or 1-888-793-WELL), the Susan G. Komen Foundation (www.komen.org or 1-800-462-9273), Gilda's Club (www.gildasclub.org or 917-305-1200), CancerCare, Inc. (www.cancercare.org or 1-800-813-HOPE), or the helpline of the American Psychosocial Oncology Society (www.apos-society.org or 1-866-276-7443).

Internet Sites for Children

Kids Konnected
www.kidskonnected.org

Kids helping kids is the core of this unique site. Headed by children under the direction of caring adult volunteers, it offers a number of valuable services. The hotline is available 24 hours a day, 7 days a week at 1-800-899-2866. Children can talk with someone who can answer questions, share concerns, or just listen. Children can log on and chat in the chatroom, sign up for the Kids Konnected newsletter, check out the up-to-date database on books about cancer and coping skills, and find out about events and support groups in their local area. (Note: Kids Konnected never gives medical advice.)

APPENDIX 4

Annotated Bibliography

Books for Children

The Hope Tree: Kids Talk About Breast Cancer (ages 4–8)
Laura Numeroff and Wendy Harpham, M.D.; David McPhail, illustrator
Simon & Schuster, 2000

Ten important topics to families dealing with *any* type of cancer are presented by a fictional support group made up of a variety of animals of different ages. Vignettes and advice help children accept what's happening, find ways to help their situation, and nourish hope. Charming illustrations. All authors' proceeds donated to the Susan G. Komen Breast Cancer Foundation.

Mommy's in the Hospital Again (Ages 4–7)
Carolyn Stearns Parkinson; Elaine Verstraete, illustrator
Solace Publishing, Inc., 1996

For families dealing with chronic illness and a parent's recurrent hospitalizations, this book discusses the importance of enjoying happy times even when a parent is ill. The difficulty of balancing conflicting emotions is presented in a realistic yet gentle and hopeful way.

Kemo Shark (Ages 4–12)
Kidscope
Mass Market Paperback, 1995
English or Spanish
404-892-1437 to order free copy
www.kidscope.org to order free copy or to download

A color "comic book" designed to help children understand the changes in a parent who is undergoing chemotherapy.

Because Someone I Love Has Cancer: Kids' Activity Book (Ages 4–8)
American Cancer Society, 2002
Spiral-bound activity book with colored markers.
1-800-ACS-2345
www.cancer.org

What's Happening to Mom?
Susan G. Komen Foundation, 2002
1-800-462-9273 (1-800-I'M AWARE)
www.komen.org

This free booklet provides information on helping children understand their mother's diagnosis of breast cancer as well as information about how the patient can solicit support from her children. English and Spanish versions of the booklet are available.

The Paper Chain (Ages 4–8)
Claire Blake, Eliza Blanchard, and Kath Parkinson
Health Press, 1998

Lovely watercolors accompany this simple story that provides hope along with explanations of surgery, chemotherapy, and radiation.

Our Mom Has Cancer (Ages 9–12)
Abigail Ackermann and Adrienne Ackermann
American Cancer Society, 2002
www.cancer.org

Two sisters, ages 11 and 13, describe what it was like for them when their mom was diagnosed with breast cancer and underwent surgery and chemotherapy.

My Family Is Living with Cancer (Ages 6–10)
Sandra Peyser Hazouri and Miriam Smith McLaughlin
Marco Products
www.marcoproducts.com

A story for parents and children to read together. Includes helpful suggestions for parents, such as how to help children deal with a parent's hospitalization.

What Is Cancer Anyway? Explaining Cancer to Children of All Ages (Ages 4–9)
Barkey and Eve Series, Book 5
Karen L. Carney
Dragonfly Publishing, 1998

Two lovable main characters define cancer and explain cancer treatments in hopeful ways. Drawings lend themselves to being colored by readers.

My Mommy's Cancer (Ages 9–12)
My Daddy's Cancer
Cindy Klein Cohen, M.S., C.C.L.S., and John T. Heiney; Michael Gordon, illustrator
Promise Publications, 1999

These two interactive stories contain activities that encourage children to express their feelings. They also contain a guide for parents.

Our Family Has Cancer, Too! (Ages 7–12)
Christine Clifford; Jack Lindstrom, illustrator
Pfeifer-Hamilton Publishers, 1997

This small book tells the story of a mom going through cancer treatments. What makes the book special are the cartoon pictures, workbook pages, and suggestions for family discussions.

When Someone Has a Very Serious Illness: Children Can Learn to Cope with Loss and Change (Ages 9–12)
Marge Heegaard
Woodland Press, 1992

A workbook to help children deal with their feeling about serious illness and find ways to cope.

Books for Parents

How to Help Children Through a Parent's Serious Illness
Kathleen McCue
St. Martin's Press, 1994

A gentle and practical guide, written by a child-life specialist. This book is a must-read when a parent's treatment requires hospitalization.

Cancer in the Family: Helping Children Cope with a Parent's Illness
Sue P. Heiney, Joan F. Hermann, Katherine V. Bruss, and Joy L. Fincannon
American Cancer Society, 2001

A guide to helping young or adolescent children deal with a parent's diagnosis of cancer. Includes a separate workbook to help young children to express their feelings through art.

Can I Still Kiss You?: Answering Your Children's Questions About Cancer
Neil Russell
Health Communications, 2001

Russell discusses the effect of his cancer diagnosis, surgery, radiation, and chemotherapy on himself and his family. *Can I Still Kiss You?* is a blend of narrative and interactive journal for parents and children.

Videos for Parents

We Can Cope: When a Parent Has Cancer
Inflexxion, Inc.
320 Needham Street, Suite 100
Newton, MA 02464
Phone: 617-332-6028
Toll free: 800-848-3895
Fax: 617-332-1820
www.wecancope.com

This Parent Video is hosted by Dr. Wendy S. Harpham and provides parents with suggestions for finding both the words and the hope to deal with the difficult task of talking with their children about cancer. Inflexxion, Inc.,™ the producer, also offers a Child Video and a Teen Video.

Talking About Your Cancer: A Parent's Guide to Helping Your Children Cope
Fox Chase Center
Department of Social Work Services
8801 Burholme Avenue
Philadelphia, PA 19111
Phone: 215-728-2668 (ask for Marge Winters)

This 18-minute video shows parents how to break the news and take those first painful steps toward surviving as a family. Short dramatization sequences by professional actors are artfully meshed between advice given by former Surgeon General C. Everett Koop, M.D., and real-life mothers, fathers, and children who have been there.

My Mom Has Breast Cancer
Produced by Kidscope
Available free of charge through the Community Service Section of many Blockbuster Video stores, many public libraries, or www.kidscope.org

Interviews with seven children and four mothers who have successfully gotten through the mothers' experiences with breast cancer treatment. Various ways of coping are described.